Travis Mesonum
Marathon Texas July 4/2005

Texas Tales of LOST MINES and BURIED TREASURES

Carlos Savoy

Republic of Texas Press

Library of Congress Cataloging-in-Publication Data

Savoy, Carlos.
Texas tales of lost mines and buried treasures / Carlos Savoy.
p. cm.
Includes bibliographical references (p.) and index.
ISBN 1-55622-878-3 (pbk.)
1. Texas—History, Local—Anecdotes. 2. Texas—Antiquities—Anecdotes. 3. Treasure-trove—Texas—Anecdotes. 4. Mines and mineral resources—Texas—History—Anecdotes. I. Title.

F386.6 S28 2001
976.4—dc21 2001041904

Republic of Texas Press is an imprint of Wordware Publishing, Inc.

Printed in the United States of America

ISBN 1-55622-878-3
10 9 8 7 6 5 4 3 2 1
0110

All inquiries for volume purchases of this book should be addressed to Wordware Publishing, Inc., at 2320 Los Rios Boulevard, Plano, Texas 75074. Telephone inquiries may be made by calling:

(972) 423-0090

Contents

Introduction

Not long ago on the Caribbean island of Jamaica, approximately one dozen of the world's most famous and successful professional treasure hunters gathered to discuss ongoing projects, to reminisce about previous expeditions and successes, and to discuss future plans and projects.

During the gathering, three prominent adventurers were engaged in a discussion about the most promising land- based sites in the world for locating lost mines and buried treasures. The three men were: Will Sexton, noted British treasure hunter who conducts most of his expeditions in the United States, Mexico, and Central America; Quinton McClain, a New Zealander whose exploits and successes in Africa are legendary among his peers; W.C. Jameson, the renegade American treasure and artifact hunter and best-selling author of books on lost mines and buried treasures.

These three treasure hunters ultimately concluded that of all the countries in the world, the United States possesses the greatest number of real and potential sites for lost and hidden treasures. Furthermore, these three intrepid outdoorsmen agreed that, of the entire U.S., the state of Texas likely held more of these locations than any other.

Most serious treasure hunters, as well as those who delight in the tales, legends, and myths of lost mines and buried treasures, would likely agree. To be sure, Texas is, indeed, the source of some of the world's most fascinating and compelling history regarding such stories.

As a result of a near lifetime of searching for and occasionally finding lost mines and buried treasures throughout the state of Texas, along with decades of collecting information and materials from a number of other treasure hunters, authors, researchers, and archivists, I have accumulated several file cabinets filled with such stories, documentation, and interviews.

This book represents only a small portion of that collection but includes most of my favorite Texas tales. In the pages that follow, I am proud to share with the reader some of the stories that have captivated residents of the Lone Star State and others for generations, stories that have been the source of constant inspiration relative to hunting for lost mines and buried treasures, and stories and tales that have been responsible for initiating some of the greatest adventures of my life.

What I like best about these stories, however, is that in virtually every case, the treasure is still there, still lost or hidden, still waiting to be found.

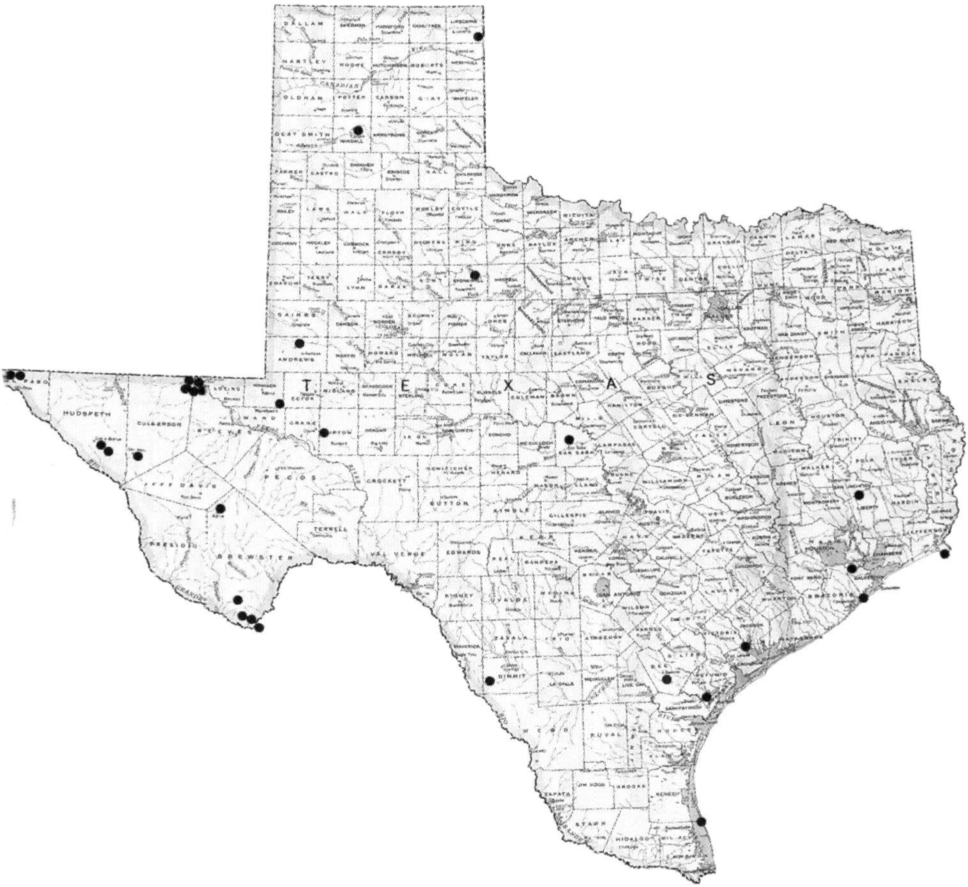

The approximate locations of lost mines and buried treasures.

The Guadalupe Mountains

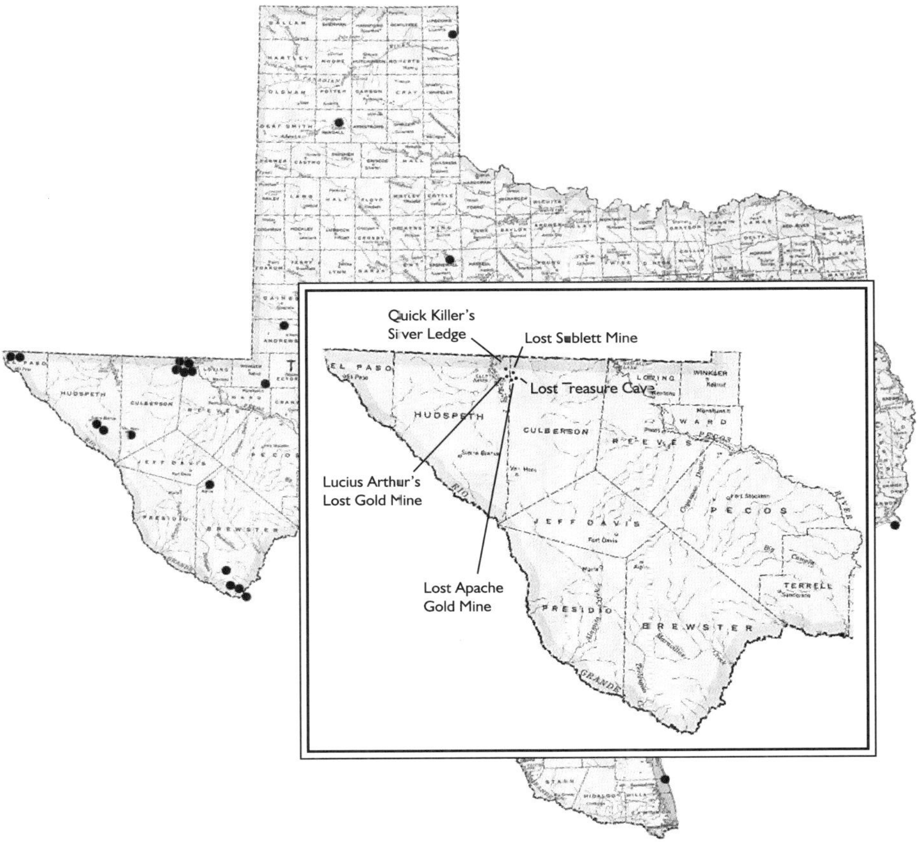

Rising from the sandy and gravelly plains of the northern Chihuahuan Desert in West Texas are the Guadalupe Mountains. The range, which includes the tallest peak in Texas—El Capitan at 8,749 feet—is extremely rugged, relatively remote, and even today remains a serious challenge to experienced hikers, backpackers, and seekers of lost mines and buried treasures.

The Guadalupe Mountains are located, for the most part, in Culberson County in West Texas, with some of the range spilling across the state line into New Mexico to the north. This high, broken, and impressive range is generally composed of limestone formed from fossilized algae that was laid down in an ancient sea beginning some 500 million years ago. Ongoing series of deposition, uplifts, and earthquakes, accompanied by millions of years of erosion from flowing water and wind, have sculpted this picturesque range to its present appearance.

So rugged and forbidding are the Guadalupe Mountains that the range became the easily defendable homeland for the Mescalero Apache, a wide-ranging and much feared branch of what is arguably the fiercest of all North American Indians. The Mescaleros got their name from the mescal plant, which grows throughout the Guadalupe Mountains and served as a staple food for these same Apaches. It was in the Guadalupe Mountains that the Mescaleros repelled repeated attempts by the United States cavalry to conquer and relocate them. The Mescaleros eventually succumbed to vastly superior forces and weaponry in 1869. Despite their subsequent removal from the range and onto reservations in New Mexico, many believed that one or two dozen Mescaleros remained living in the somewhat remote Guadalupe Mountains as late as the 1930s.

In 1858 the Butterfield Overland Mail opened a stagecoach station in the foothills near El Capitan Peak. This station, called The Pinery, remained in operation for only eleven months before it was abandoned in favor of a more southerly route. The initial trail, laid out by surveyors for the Overland Mail, provided a relatively simple route for travelers into and through this region.

Since the arrival of the Spanish in this part of North America during the sixteenth century, the Guadalupe Mountain range was often visited by these explorers. Legend holds that the Spaniards who came to the area arrived in search of gold and found it here in impressive quantities. To this day, Spanish artifacts dating from that period are occasionally found in the Guadalupe Mountains.

In time newcomers arrived, though never in great numbers. The aridity and the harsh environment, accompanied by the early threat from Indians as well as perceived threats from bears, mountain lions, wolves, and other kinds of wildlife, lured only a few of the most hardy types of settlers. During the twentieth century, oil and gas was discovered in the adjacent basin and plains—a former sea bottom—and the population grew slightly.

Even today the area in and around the Guadalupe Mountains is sparsely settled. Though much of the range has come under the management of the United States National Park Service, the numbers of hikers and backpackers who climb the steep slopes and enter the high country remain comparatively few.

Though it has long served as a setting for some remarkable Indian, Spanish, cavalry, and other history, the Guadalupe Mountains today are more widely known as the setting for some of the most fascinating legend and lore as they relate to lost mines and buried treasures. Beginning with the adventures of the early Spanish exploration and continuing until relatively recent times with foothill residents, the Guadalupe Mountains have become well known worldwide as a geographic region likely containing more lost mines and buried treasures than almost anyplace else on earth.

Quick Killer's Silver Ledge

The tales of gold and silver in abundance in the Guadalupe Mountains began four centuries ago with the coming of the Spanish explorers. Former New Mexico governor Lew Wallace wrote that he once discovered a document that relates this specific information. The parchment, found among a quantity of papers left over from Spanish rule in the old Santa Fe Palacio, told of how a group of friendly Indians led a Spanish officer, one Capitan de Gavilan, to a secret location in the Guadalupe Mountains where they excavated great quantities of gold nuggets and wire gold. The Spaniards, according to the document, mined this gold for several months until the great Pueblo Indian uprising of 1680 forced them to abandon the region. The legend that eventually evolved relates that the Spaniards filled in and covered the mines and removed all evidence of such labors. They intended to return and reopen them when the Indians were finally vanquished. As a result of the ever-changing political situations in Spain as well as in the United States, however, they were never able to make the journey back to the Guadalupe Mountains.

During the middle of the nineteenth century, the U.S. government sent the army into portions of the western wilderness for a variety of reasons: protecting settlers from warring Indians, escorting the men surveying for railroads, protecting travelers.

In 1863 U.S. Army Major John C. Cremony was placed in charge of a contingent of cavalry that patrolled vast portions of southeastern New Mexico and West Texas. One of Cremony's scouts was an Indian named Quick Killer. Quick Killer, described by the soldiers as very efficient and loyal, sported a long braid of black hair that hung down his back almost to his hips. The braid was ornamented with a number of large silver conchos. One day when Cremony asked the Indian about the conchos, Quick Killer told him they were made from silver he dug from a ledge located on the side of a steep canyon in the Guadalupe Mountains.

Quick Killer then related a most amazing story. He told Cremony that once, many years ago, when he and several companions were fleeing from a band of Comanches, they escaped into the Guadalupe Mountains. While climbing one canyon wall, he related to the major, he grabbed a bush with which he hoped to pull himself up. The bush gave way, exposing lumps of silver underneath. Later he returned to the location several times to retrieve pieces of the silver, which he used to fashion the ornaments he proudly displayed in his hair.

Several months later while on patrol in the foothills of the Guadalupe Mountains, Quick Killer rode up to Major Cremony and told him they were not very far from the silver he spoke of earlier. If the major wished, said the Indian, he would take him to it in the morning.

The next morning Quick Killer led Cremony into a canyon where the two men dismounted and tied their horses to a tree. From this point, they climbed approximately 300 feet up one side of a steep canyon wall until, according to Cremony, they reached "a bold and unmistakable mineral ledge, thickly shrouded with underbrush and stunted trees." After examining the location for a few moments, Quick Killer walked to a certain spot, dug a few inches into the ground, and pulled out several large pieces of silver, which he handed to the major. Cremony later wrote that he was very impressed with the quantity and quality of the silver found that day.

Cremony's duties eventually took him elsewhere, and there is no record that he ever returned to the Guadalupe Mountains to recover more of the silver. To date, no record has ever been found relative to the fate of Quick Killer. Thus, the location of the silver ore in the Guadalupe Mountains remains a mystery.

Cremony's entry in his journal of the silver found in the Guadalupe Mountains may have been the first formal reference to the likelihood that precious metal could be found in the range. Since then, mention of gold and silver being discovered in these mountains has become quite common.

When white settlers began moving into the American Southwest, they often came in contact with the native tribes who lived there. From the time the earliest Anglo settlers arrived in Texas, New Mexico, and even Arizona, they noted that, while the Apaches had little appreciation for the immense wealth associated with gold, they noticed that the Indians often wore jewelry and armbands made of gold and silver. The Indians also used gold from time to time to purchase rifles and ammunition. When pressed for the source of their ore, the Apaches always responded that it came from the Guadalupe Mountains in Texas. Once, the famous Apache leader Geronimo was asked where his tribe's gold came from, and, according to folklorist J. Frank Dobie, he replied that "the richest gold mines in the western world lay hidden in the Guadalupes."

Noted Apache leader Mangas Coloradas arrived at a supply store in Fort Sumner, New Mexico, to purchase rifles and ammunition. He paid for the goods with large nuggets of gold. When asked by the storekeeper where he obtained the nuggets, the Apache replied that they came from the Guadalupe Mountains.

A man named George Bledsoe arrived in El Paso, Texas, sometime in the 1880s with a quantity of gold he wanted assayed. When informed that the gold was almost pure and incredibly rich, he mentioned that it came from a location in the Guadalupe Mountains.

In 1896 an old Mexican told of how he was taken captive by Mescalero Apaches and forced to live with them in the Guadalupe Mountains. One day, he recalled, while accompanying his captors on a hunting trip, they paused in a dry streambed to pick up several large nuggets of gold.

Green Ussery, a West Texas rancher, was riding his horse up a canyon in the Guadalupe Mountains looking for some stray cattle when he discovered a large gold nugget in the streambed.

When Quick Killer led Major Cremony to the silver deposit in the Guadalupe Mountains, he little realized that he was introducing the notion that this vast range possibly served as a setting for great deposits of rich ore.

The Lost Apache Gold Mine

Despite the slight breeze generated by a small window fan, the second-floor chiropractor's office remained hot, humid, and quite uncomfortable. Glancing out the window around 10:00 a.m., the chiropractor noticed hardly anyone moving on the streets and sidewalks. Even the pigeons were noticeably absent from the nearby tree-lined plaza. The Texas summer of 1926 would go down as one of the hottest on record.

Presently the chiropractor observed an aged man making his way along the sidewalk, keeping close to the narrow strip of shade offered by late morning shadows. Now and then the old man would stop to read an address, consult a small, crumpled piece of paper he held in his weathered hand, then move on. With the passage of a few more minutes, the old man finally paused just below the window from which the chiropractor watched. After tucking the piece of paper in a pocket, the old man opened a door and entered the building.

A short time later the chiropractor could hear the old man talking to the receptionist in the adjoining room. In halting and broken English, he told the receptionist his name was Polycarpio Gonzalez, that he was deaf, and that his daughter suggested this might be a place to find some help with regaining his hearing. Moments later the old man was led into the chiropractor's office. The chiropractor guessed the new client to be in his seventies, perhaps older. Trying to appear nonjudgmental, the chiropractor noted the old and ragged clothes worn by the elderly man and wondered to himself how he would be able to pay for any treatment he might receive.

Using a combination of his poor English and hand signs, Polycarpio Gonzalez explained to the chiropractor that he wanted more than anything to hear the laughter of his great-grandchildren. In addition, he said, he wanted to be able to hear music again. He loved music, he said, and had been unable to hear any for the past two decades.

When Gonzalez finished, the chiropractor gave him a brief examination and told him he would try to help with a series of treatments. After escorting the old man back out into the receptionist's office, he scheduled Gonzalez for weekly visits, bid him goodbye, and returned to his office.

Several weeks later, as Polycarpio Gonzalez was leaving the chiropractor's office following his fourth visit, he politely asked for permission to sit for a while in the waiting room. While he was relaxing in a chair and looking at photographs in a magazine, the receptionist turned on a radio and tuned it to a music station. As the sound flooded the room, Gonzalez looked up and his eyes widened. Jumping up from his chair, he began shouting in Spanish that he could hear the music! As the chiropractor came out of his office to investigate the source of the excitement, he saw the old man dancing across the floor in joy. On seeing the chiropractor, Gonzalez approached him, embraced him, and thanked him over and over for helping him get his hearing back.

After giving Gonzalez time to catch his breath, the chiropractor asked him about the curious dance steps he just witnessed, stating he had never seen anything like them before. Placing an old, weathered hand on the chiropractor's shoulder, Gonzalez told him he learned the dance from the Mescalero Apaches when he was a boy. Before the chiropractor could inquire further, Gonzalez said he must go home and tell his daughter the good news.

One week to the day later, Gonzalez returned to the chiropractor's office and asked to meet with him. After closing the door, the chiropractor inquired about the old man's hearing. Gonzalez assured him everything was fine, but that he wanted to speak of another topic. He explained to the chiropractor that he was so happy now to be able to hear music that he sat in front of the radio for hours at a time listening to the wonderful sounds emanating from the little box. His eyes moistened as he spoke of how much he loved to hear the stories told to him by his great-grandchildren. He told of being able to hear other things that he had missed for so many years. He concluded by thanking the chiropractor profusely and explaining he had no money to pay for his treatment.

Instead, Gonzalez told the most incredible story of a lost gold mine the chiropractor had ever heard.

When Polycarpio Gonzalez was approximately ten years old, he was informally adopted by a United States Army cavalry officer named Colonel Boone. Boone was given the responsibility of protecting new settlers and occasional travelers from warring Apaches, and he led a contingent of troops stationed at Fort Stockton, Texas. The young boy's origins remain a mystery, but it is believed his father was an Apache and his mother a

Mexican. The Indian father was killed in a skirmish with Boone's own soldiers, and the boy was given to the officer by his mother because she was no longer able to care for him and his three siblings.

The boy often accompanied the soldiers on various missions and explorations, his ability to scout and follow a trail proving quite valuable to the troopers for one so young. During an extended expedition that carried Boone and a dozen cavalrymen to the Rio Grande border, the soldiers were preparing to make camp in a narrow, shallow canyon when they were attacked by approximately fifty Mescalero Apaches. A brief but violent battle ensued, ending when all of the soldiers were killed. Young Gonzalez was badly wounded in the thigh and was eventually found hiding among a small cluster of bushes. After being dragged from his place of concealment, he was immediately recognized as being the nephew of one of the warriors. Gonzalez' wound was treated, and he left with the Apaches, who now considered him a member of the tribe.

Gonzalez remained with the Mescalero Apaches for the next eight years. They taught him the ways of the tribe, and in time he became a loyal and reliable member of the band. In addition he became a competent warrior, often engaging in battle with the U.S. Cavalry and taking scalps on several occasions. While living with the Indians, he learned the Mescalero language, he learned to live off the land the Apache way, he learned their songs, and he learned their dances.

A principal campsite for the Mescalero Apaches was Manzanita Spring located near the bottom of the southeast-facing slope of the escarpment. Here, deer and elk were plentiful and the waters of the spring were clear and cool. While the men brought in game and conducted raids on nearby ranches, the women harvested and prepared the abundant mescal plant. All things considered, life here was good for the Mescalero Apaches.

One evening Gonzalez was asked to meet with two of the tribe's elders. He was told that only a few minutes' hike from the spring was the source of all the gold that the Apaches used to make ornaments and trade for rifles and ammunition. At dawn the following morning, Gonzalez was invited to accompany one of the elders to the gold mine to help gather some of the precious metal to trade for new weapons. The two men, one aged and one growing into a respected warrior, walked single file along the narrow game trail that paralleled the rim of a shallow canyon that had been cut into the ground by the occasional water that rushed down the slopes from the mountains following a rain. At one point along the trail, the elder stopped and indicated a number of stacked rocks at the bottom of the canyon near the far wall. Behind those rocks, said the elder, was the entrance to the famous Apache gold mine.

The two made their way to the bottom of the rocky, boulder-strewn canyon and were soon standing before the concealed shaft. At a signal from the elder, they began removing the large rocks that covered the opening. Once the tunnel was exposed, the elder retrieved a grass rope from its hiding place behind some rocks and tied one end of it to a nearby juniper. After lighting a torch, he used the rope to descend into the steep, dark shaft. Gonzalez followed. The shaft was just over fifty feet deep, and once at the bottom, the young warrior marveled at the thick seams of shiny gold woven throughout the quartz rock. Some of the seams were as wide as the span of his hand.

The elder showed Gonzalez how to carve the soft metal from the seams.

After placing several pieces of the gold into a leather belt-pouch, the elder announced it was time to leave. After exiting the shaft, they replaced the rocks that covered it, returned the grass rope to its hiding place, and departed for camp.

The following morning a Mescalero scout rode into the camp and excitedly informed the leader of an approaching cavalry unit. Though the soldiers were still two days away, the chief summoned his followers, told them they would have to leave, and instructed them to begin packing.

As the Indians started loading goods onto horses and mules, the chief told Gonzalez and two other young warriors to run to the gold mine and conceal the entrance such that the white men could never find it. Hurriedly, the three rolled and toted large rocks to the shaft and threw them in, almost completely filling the tunnel. The entrance was then covered with more rocks. By the time Gonzalez and his companions were finished, the entrance to the shaft was undistinguishable from the rest of the canyon.

That evening, as the Mescaleros prepared to leave, the chief called Gonzalez aside, informed him he was leading the tribe into Mexico, and that it would be a good time for him to return to his people. Several minutes later Gonzalez was back at the now abandoned campsite, watching as his family and friends rode away to the southwest. The year was 1877.

When Polycarpio Gonzalez finished his tale, he told the chiropractor that, in payment for curing his deafness, he would take him to the Guadalupe Mountains and show him the location of the old Mescalero gold mine.

Intrigued by the old man's story and excited at the possibilities of locating a fortune in gold in the Guadalupe Mountains, the chiropractor began making arrangements for travel. Within two weeks he had located a colleague who agreed to take over his practice until he returned, purchased camping supplies as well as some digging tools, and borrowed an automobile capable of making the long trip west.

Days later the chiropractor, accompanied by Gonzalez, arrived in Pine Springs, a settlement comprised of two or three families located in the shadow of El Capitan Peak, a prominent mountain in the Guadalupe range and one of the tallest peaks in Texas. Here, the automobile was left with a rancher, and the chiropractor arranged for the loan of two stout horses and a pair of mules to transport himself, Gonzalez, and the supplies to the old Mescalero campground at Manzanita Spring, located almost two miles away across rugged terrain. Just before sundown the two men set up camp only a few feet from the pool of water. As the chiropractor prepared an evening meal, Gonzalez looked around the old site, constantly turning and regarding the adjacent mountains and plains. Memories of when he lived here with the Mescalero Apaches some fifty years earlier flooded his thoughts.

The following morning after breakfast, the two men followed a well-used game trail that paralleled the dry wash that ran just east of the campsite. After several minutes of hiking, Gonzalez halted and pointed to a jumble of rocks next to the opposite wall of the wash. Carefully, the two slid down the steep slope to the bottom. After leading the chiropractor over to the pile of rocks, Gonzalez fell deep in thought for several minutes, then finally turned toward his companion and announced that beneath the small boulders lay the shaft that contained the gold of the Mescaleros.

Immediately the two men began laboring to remove the heavy rocks. The work was hard and was made even more difficult due to the advanced age of Gonzalez and the fact that the chiropractor was unused to such labor. After several hours and numerous rests, however, the two men eventually uncovered the opening to a shaft.

The chiropractor and Gonzalez continued to work for the remainder of the day. By the time they finally quit, they had pulled, pushed, and hauled enough of the heavy stones to expose about five feet of the old shaft. Just inside they found very old juniper logs positioned to shore up the crumbling interior.

The next morning when they returned to the site of the mine, the chiropractor determined that it would be extremely difficult to remove any more rocks from the shaft without some extra help and the appropriate equipment. They must go back to San Antonio, he said, and employ some workers.

After returning to San Antonio, the chiropractor experienced very little luck in attempting to find men and equipment necessary for removing the huge stones from the mineshaft. The delays continued for several months and then the chiropractor received some very bad news: Polycarpio Gonzalez had died in his sleep.

A year passed, and the chiropractor was still unable to lure workers to the Guadalupe Mountains of West Texas. Closing down his business once again for a time, he traveled to El Paso, some 100 miles west of the

Guadalupe Mountains. Here, he was convinced, he would have better luck securing men and equipment for the daunting task of pulling the rocks out of the shaft.

It was not to be. After spending several weeks in the border town and having no success, the chiropractor, discouraged and disappointed, finally drove back to San Antonio. Several years passed, and he eventually gave up any hope of locating the gold of the Mescaleros.

Today, trails lead to the rim of the wash in which Polycarpio Gonzalez and the chiropractor found the shaft containing the gold. Decades have passed since the two men removed the stones from the opening, and dozens, if not hundreds, of rainstorms in the Guadalupe Mountains and along the adjacent slopes have generated swiftly flowing runoff, pouring thousands of gallons of water down the many channels stretching from the slopes across the foothills toward the plains to the south and southeast. As these flash floods course through the washes, they carry with them dense loads of fine gravel and sand eroded from upstream and deposit them here and there along the twists and turns of the canyon. In this manner, the secret gold mine of the Mescalero Apaches was likely filled in and covered over such that it looks much like the rest of the canyon today. Perhaps other flash floods may uncover the shaft, perhaps not.

Though others have searched for the lost gold mine of the Mescalero Apaches, it has never been found. It lies there today next to the floor of the rocky wash and only one mile from the El Paso-Carlsbad highway, a busy road that carries travelers and goods across the countryside.

Lucius Arthur's Lost Gold Mine

The lost gold mine of the Apaches described by Polycarpio Gonzalez may have actually been found a short time after the Mescaleros abandoned their Manzanita Spring campground and then lost again. Lucius Arthur, who went by the nickname Frenchy, once related a story about a lost gold mine for which people have been searching to this day.

Arthur's story was told to a West Texas rancher named Hardesty. Hardesty, in turn, related the account to others, and today this tale is part of the legend and lore of this grand mountain range.

For a time, Arthur lived in Ysleta, Texas, about twenty miles downriver on the Rio Grande from El Paso. While eating dinner in a tavern one evening, he overheard a conversation between two Mexican men about a gold mine. While he pretended to be occupied by his meal, Arthur listened intently to the two men as they spoke of a shaft located at the bottom of a canyon or wash in the Guadalupe Mountains that apparently yielded incredible quantities of gold. As he listened, Arthur learned that the two men planned on departing at dawn of the following day to return to the mountain range and retrieve more of the gold.

Arthur decided he would follow the two men and learn for himself the location of this rich mine. The next morning Arthur hastily packed some food and water while he watched and waited for the two men to depart Ysleta. When the Mexicans rode out of the small town, each of them leading a mule, Arthur followed them, staying about one mile behind them so he would not be seen. By the end of the second day of travel, Arthur had run out of food and water and saw no immediate prospect of attaining more. Frustrated at his poor planning, he returned to Ysleta.

While he waited for another opportunity to trail the two Mexicans to the location of their gold mine, Arthur discovered that one of them was the son of a wealthy hacendado who lived in Mexico. Years earlier, the family made an annual journey to the large playa, or salt lake, located just to the south and west of the Guadalupe Mountains where they would load the fine salt they found there into a large wagon and return to the rancho. During one of these trips to obtain salt, as the story was related, two members of the family rode into the Guadalupe Mountains to hunt for game. Before they were able to shoot a deer, they discovered a rich deposit of gold. Each year after that, the family not only came to the area for salt, but invested several days in digging the gold from the source.

Approximately two months later, Arthur watched for the departure of the two Mexicans. This time when they rode out, Arthur had equipped himself with a full week's worth of provisions. For four days Arthur tracked the two men—each riding a horse and leading a mule, one of which was carrying supplies—across the arid desert until arriving in the foothills of the Guadalupe Mountains just before sunset of the last day. After winding along an old, narrow trail, the two men stopped near the lip of a narrow arroyo, dismounted, and set up camp as Arthur watched from a safe distance. Arthur's only concern at the moment was that no water for himself or his horse appeared to be available anywhere.

The next morning the two men rose and prepared breakfast. After eating, one of them unrolled a rope ladder, secured one end to a stout juniper tree that grew at the edge of the arroyo, and dropped the rest of it into the deep, narrow gulch. After lowering some digging tools into the chasm by rope, the two men descended the ladder.

Arthur grew cramped, tired, and anxious as he waited in his hiding place. After about four hours, one of the men finally climbed out of the chasm with a leather bag filled with water and a smaller bag—an ore sack—which he deposited near the juniper tree. After watering the horses and mules, the man took a long drink for himself, set the bag down, and returned down the rope ladder.

During the long afternoon, Arthur watched as several more small bags were brought up and set next to the first one. Though he had no way of knowing, Arthur assumed the bags contained gold ore.

On the morning of the fourth day, the two Mexicans loaded the ore sacks into heavy leather saddle packs and tied them to the mules. After packing their gear and saddling their horses, they rode back down the long trail toward Ysleta.

Once the two men were several miles away, Arthur left his hiding place and rode over to their campsite to inspect the arroyo. Arthur later related that the chasm extended from the base of the Guadalupe Mountain escarpment downslope as far as he could see, with the width varying from approximately 40 to 100 feet. From where he stood on the lip of the steep wall, the arroyo was approximately 60 feet deep. At the base of the opposite wall, Arthur noted the entrance to what was apparently a mineshaft, an opening that had a number of small boulders scattered and lying around the front of it.

On the one hand, Arthur was excited at the discovery of what he believed to be a very rich gold mine. On the other hand, he was disappointed—unaware of the characteristics of the site, he had brought with him nothing that could serve as a rope ladder to descend to the bottom of the narrow arroyo.

Arthur decided he would return immediately to Ysleta where he would obtain a rope ladder and enough provisions to last for at least two weeks. As he considered the great wealth he knew he would soon obtain, he looked around the campsite. Here and there he found small pieces of quartz, apparently dropped by the Mexicans. Picking one of them up, he discovered that it contained an amount of gold. In all, he found about a dozen of them and placed them in a shirt pocket.

On returning to Ysleta, Arthur went about securing the necessary equipment to make another trip to the gold mine in the Guadalupe Mountains. On the evening before his departure, he confided in a close friend and showed him the gold-laced pieces of quartz. The friend cautioned him about making the trip alone and warned him that hostile Mescalero Apaches still lived in and around the Guadalupe Mountains. Arthur told his friend he was convinced the Indians were not a threat.

The next morning Arthur rode out of Ysleta leading a newly purchased mule. He was never seen again.

Whatever happened to Lucius Arthur remains a mystery to this day. Whether or not he ever returned to the mysterious arroyo and mined some of the gold likewise remains a mystery.

Most who lived in the area and were acquainted with Lucius Arthur were convinced he was discovered by the Mescalero Apaches and killed. Others have offered the suggestion that he may have struck it rich and rode away, simply choosing not to return to Ysleta.

Some who have spent near lifetimes researching tales of lost mines and buried treasures in the Guadalupe Mountains are convinced that the gold mine described by Lucius Arthur was the same one to which Polycarpio Gonzalez led the chiropractor. Some few are also convinced it may be the same location for the source of one of the most famous lost mine legends of all time: The Lost Sublett Mine.

The Lost Sublett Mine

Without question, the most famous tale of lost gold in the Guadalupe Mountains is one most people refer to as the "Lost Sublett Mine." This mine, if it exists, and it likely does, has been in the thoughts of treasure hunters, professional and amateur alike, for well over a century. Over time, countless hours, energy, and money have been invested in the search for the Lost Sublett Mine. Even today hopeful seekers of lost treasure enter the Guadalupe range believing they will be the one to finally locate the source of Old Ben Sublett's lost gold.

Sublett came to the Guadalupe Mountains via a roundabout route. Born William Caldwell Sublett in Kentucky or Tennessee, he left home at an early age, eventually winding up in Missouri and marrying Laura Louise Denney. Work was scarce in Missouri, and the two traveled to Colorado hoping to find employment in the mines, but it was not to happen. Following several more months of travel in a wagon pulled by a pair of scrawny mules, the two finally ran out of money in Monahans, Texas, where they lived in a tent on the outskirts of town.

While Laura took in washing in an attempt to make some money, Sublett made his way through the small town looking for any kind of work that might be available. Now and again he would secure a job as a house painter or loading heavy bundles at the local mercantile, but more often than not these jobs didn't last long. Part of the reason Sublett couldn't keep a steady job was that once he received his pay, he spent a large portion of it on drink. It got so bad that he sometimes showed up for work in a drunken stupor and was generally sent away. Eventually word got out about Sublett's habits and jobs grew scarce.

Since Sublett had very little money, he could not afford to purchase new clothes and shoes when his old ones became worn. He was often seen around town dressed in little more than rags. Because of his appearance, along with his reluctance or inability to bathe, he was often shunned by the town's citizens. When employment became difficult to find, Sublett spent most of his time in town begging for handouts.

During idle times when he was not employed, Sublett hitched his poor mules to his rickety wagon and traveled to the Guadalupe Mountains some 125 miles to the west where he prospected for gold. Time and again he was cautioned about hostile Apaches living in and traveling through the range, but Sublett ignored the warnings. As a result of his indifference to the potential Indian menace, his inability to hold a job, his occasional bouts of drunkenness, his disheveled appearance, and his increasingly eccentric behavior, Sublett earned a reputation of being crazy. In time he was simply referred to as "Old Ben" and was regarded as demented from drinking and spending too much time in the sun.

Though things never got much better for Old Ben and Laura, they managed to have three children—two girls and a boy. Laura Sublett was concerned that the children had no opportunities for school in Monahans. She was also growing concerned that the town was becoming too wild with whiskey peddlers and gamblers and was not a fit place to raise children. Laura pleaded with Old Ben to move the family to Odessa, some thirty-five miles to the northeast. She was also convinced that Ben would be more likely to find work in the larger town. Unfortunately, the reputation Old Ben earned in Monahans followed him to Odessa, and he found getting a job there just as difficult, if not more so.

Several months following the move to Odessa, Laura died from complications related to tuberculosis, and Old Ben was faced with the prospect of taking care of his three children. The oldest daughter continued to wash clothes for the townspeople while watching over her younger siblings. Now and again Old Ben was able to land a job for a few days, but he could find nothing permanent.

In spite of all his troubles, Sublett still found opportunities to travel to the Guadalupe Mountains to prospect for gold. Time after time he would ride out of Odessa in his old wagon pulled by the two decrepit mules, only to return days or weeks later with nothing to show for his efforts. Already considered the town fool, Old Ben became the butt of even more jokes and insults in Odessa. Once, after being thrown out of Mollie Williams' Saloon for panhandling, Old Ben told everyone he was going to strike it rich in the Guadalupes, and that when he returned to Odessa a wealthy man, he would buy everyone in town a drink.

The saloon customers merely laughed at him.

A few days later Old Ben found a job painting a building in town. He was teamed with an aged Indian, and during breaks from the work, the two would converse. When Old Ben told the Indian, a Mescalero Apache, that he sometimes traveled into the Guadalupe Mountains to prospect for gold, the Indian replied that he knew the location of a very rich deposit of gold in the range, a place where one could go and pick up large, almost pure, gold nuggets from a dry streambed. When Old Ben asked the Indian why he didn't retrieve some of the gold so he didn't have to paint buildings for a living, the Apache replied that he had been banished from the tribe years earlier, and that if he were caught in the homeland of the Mescalero, he would be tortured and killed.

At Old Ben's urging, the aged Indian provided directions to the rich placer deposit, sketching them out on the sandy ground with a stick. When the job of painting the building was completed, Old Ben hitched his mules to the wagon and departed for this new location in the Guadalupe Mountains, the first of many such trips.

His travels to the mountain range were more frequent now. He had difficulty deciphering the old Indian's directions, but he was convinced it was only a matter of time before he finally located the gold deposit.

Old Ben's absences from Odessa and his young family became a growing concern to neighbors, and eventually documents were drafted that would remove the children from Sublett's care and place them in foster homes where they could enjoy the pleasures of a normal life and not be forced to live in abject poverty.

Approximately one week after the paperwork to secure the children was begun, Old Ben Sublett returned from one of his trips to the Guadalupe Mountains. He pulled his old, creaking wagon up to the front of Mollie Williams' Saloon, swaggered inside, and dumped a leather poke of large, bright, shining gold nuggets onto the bar. As men gathered around to gaze at the size, quality, and quantity of gold lying before them, Old Ben ordered drinks for everyone. While the men were sipping their beers, Sublett told them he had just discovered "the richest gold mine in all of North America." From that day on, no one considered William Caldwell Sublett a bum in Odessa—he was now a very wealthy man.

The next morning Old Ben purchased new clothes for his children—the first they'd ever owned. He also made arrangements for them to attend school and be cared for while he made journeys into the Guadalupe Mountains to retrieve more gold. Now his trips were more comfortable, for he purchased a new wagon and two fine mules. Approximately every three weeks, Sublett would travel to the mountains, spend a few days, and return to Odessa, each time with two to three pouches filled with gold. Those who saw the nuggets described them as being unusually large. A jeweler who examined some of the ore claimed it was the purest gold in the natural state he had ever seen.

Before long, some Odessa citizens were offering to invest in Old Ben's enterprise. Many of those who earlier scorned him and made fun of him now wanted to partner with him. Old Ben politely refused their offers, reminding them that it was only a short time earlier that they were saying he was crazy.

As he expected, Old Ben was often followed when he left Odessa. Knowing he would have to deal with such tactics, Sublett employed extreme measures to discourage his trackers. Sometimes he would set out toward the mountains and camp on the Pecos River for three or four days, only to pack up and return to Odessa. On other occasions, he easily eluded those who followed him, always returning to town days later with more gold.

When Old Ben returned with gold, he always converted it to cash in Odessa and deposited it in a Midland bank owned by W.E. Connell. Midland, a town similar in size to Odessa, was located some twenty miles to the northeast. Banker Connell noticed that whenever Sublett's account was running low, the old man would make another trip to the Guadalupe Mountains. About one or two weeks later, Sublett would invariably deposit cash in his account. Connell once offered Sublett the sum of $10,000 if he would reveal to him the location of his gold. Old Ben refused with a laugh, claiming he could scoop up that much in gold in a single afternoon.

Incensed at having his proposition turned down, Connell and a partner named Gray decided to have Old Ben followed and hired a Midland cowhand named Flannigan to track Sublett to his mine. A few weeks later when Sublett's account was running low, Connell alerted Gray who, in turn, informed Flannigan to prepare to follow Old Ben into the mountains.

One morning a few days later, Old Ben left Odessa in his new wagon. After providing him with a two-hour head start, Flannigan set off behind, easily following the wagon tracks in the desert sand. For three days Flannigan had no trouble tracking Sublett through the desert, but on the fourth day, as he was following the tracks along a road that paralleled the Pecos River, he lost them. Flannigan rode ever-widening circles around the area in an attempt to pick up the tracks but was never able to find them. Frustrated, he returned to Odessa only to discover Old Ben had returned a day earlier and, strangely, had already deposited an amount of cash in Connell's bank. Where Sublett got the gold is anybody's guess, but most believe he had some cached at various locations between Odessa and the Guadalupe Mountains.

After the passage of about two years, Old Ben became friends with another elderly prospector known only as Grizzly Bill. Like Sublett years earlier, Bill was having difficulty making ends meet and was often forced to take odd jobs in Odessa to buy food.

Old Ben once told Bill that his secret mine in the Guadalupe Mountains contained more gold than he could ever use in a lifetime, and he offered to share knowledge of the location with the prospector.

Using Old Ben's directions, Grizzly Bill eventually found the rich placer mine and returned to Odessa one afternoon with a flour sack full of gold. Anxious to display his newfound wealth to Odessans, Grizzly Bill launched a celebration that lasted for two days. On the second day, he was talked into a bronc riding contest by some fellow drunks, was thrown from a wild horse, and died immediately from a broken neck.

Sublett revealed the location of his mine to another friend, one Mike Wilson. After receiving directions from Old Ben, Wilson traveled into the Guadalupes, filled several ore sacks with large gold nuggets, and returned to Odessa. Like Grizzly Bill, Wilson held a celebration, buying drinks for everyone who showed up. The party lasted for three days, with Wilson staying drunk for most of it.

After sobering up, Wilson found he had spent almost all of his gold, so he made preparations to return to the Guadalupes for more. Once back in the mountains, however, Wilson became lost and had difficulty recognizing landmarks. After several days of wandering in the range, he returned to Odessa and asked Sublett for directions once again. Old Ben, disappointed at Wilson's carelessness, refused and sent him away. Though he tried time and again, Mike Wilson was never able to relocate Sublett's lost placer mine, and years later he passed away in an old cabin in the foothills of the Guadalupes. To the last, Wilson continued searching for the lost gold mine.

There is evidence that Sublett provided directions to his gold to a man named Rufus Stewart, a sometime trail guide, hunting guide, stagecoach driver, and house painter. While serving as a guide on a hunting trip for some Texas and Pacific Railroad officials, Stewart set up camp near the Pecos River not far from the present-day town of Mentone. In the morning he would lead the hunters to antelope and deer.

That night, after the railroad men went to sleep in their tents, Stewart was cleaning up the dishes from the evening meal when he heard the approach of a wagon. Moments later the wagon pulled to a stop about twenty yards from the camp, and the driver climbed down and approached. As the newcomer neared, Stewart recognized him as Old Ben Sublett, whom he had known in Odessa. Stewart knew of Old Ben's reputation as an eccentric. He had also heard the stories of Sublett's now regionally famous gold mine. Stewart hailed Sublett and invited him to sit at the campfire and have some coffee. During the ensuing conversation, Sublett told Stewart that he was traveling to his mine in the Guadalupe Mountains to retrieve more gold. He also told Stewart that this might be his last trip to the mine as he was getting on in years. It was, he said, getting more and more difficult to climb and hike to the location and, besides, Old Ben claimed, he now had all the wealth he could possibly spend during the remainder of his life. He invited Stewart to go along with him so he could show him the location of the mine. Stewart declined the invitation, stating he had a responsibility to the hunters. Stewart also confessed he was nervous about traveling into an area inhabited by hostile Apache Indians.

The two men sat around talking until the early morning hours. When dawn broke, Old Ben told Stewart he would like to show him where the mine was located and offered to provide directions. Stewart agreed, and the two men rode several miles to the west of the camp, finally pausing on some high ground. Pointing westward, Sublett showed Stewart what he described as a "blue mound." From that point, and with the aid of a telescope, Old Ben provided Stewart with directions to the mine.

After returning to camp, Sublett bade Stewart farewell, climbed back into his wagon, and drove away toward the Guadalupes. Before leaving, he promised Stewart he would stop for another visit on his return in three days.

On the evening of the third day, Old Ben drove right up to the hunting camp and joined Stewart and the railroad men for dinner. After the men retired for the evening and only Sublett and Stewart were left by the campfire, Old Ben told Stewart he had something to show him. He spread a blanket out next to the fire and poured onto it a large quantity of gold nuggets. The nuggets gleamed brightly in the light of the fire.

Stewart was well aware of Old Ben's placer mine, but he was stunned at the size of the nuggets spread out before him. When Stewart remarked on this, Old Ben told him that the big nuggets were easier to pick up than the smaller ones, and that for every nugget lying on this blanket, he left another thousand in the dry creek bed where they lay in profusion. Sublett gave Stewart one of the large nuggets.

Sublett spent the night in camp. When he drove away in his wagon the next morning, it was the last time Rufus Stewart ever saw Old Ben Sublett.

Weeks later Stewart decided to try to find Sublett's mine. He rode to the blue mound and, using Sublett's directions, traveled into the Guadalupe Mountains. Though he searched for days, he never found the placer mine. Several more trips into the Guadalupe Mountains were likewise unsuccessful.

Only one other person was given the secret of the location of Old Ben Sublett's gold—his son, Rolth. While still a young boy, Rolth traveled into the Guadalupe Mountains with his father. He recalled being lowered to the location of the gold on a long rope, but since wealth meant little to the youngster, and because his attention was distracted by a bear cub that hung around their camp, Rolth paid little attention to the mine and its location. He came to regret his lack of attention, for as a grown man Rolth Sublett tried many times to relocate his father's rich mine, but to no avail.

William Caldwell Sublett passed away in Odessa in 1892. He had come a long way from his humble beginnings—he became a relatively wealthy man, and he ultimately provided well for his children. But by the time of his passing, his fortune had dwindled somewhat and he left little behind in the form of trusts and inheritance. What Old Ben Sublett did leave, according to professional treasure hunter and author W.C. Jameson, was an incredible legacy, one that concerns what may be among the richest lost gold mines in the history of the United States.

Jameson has searched and researched the Lost Sublett Mine more than anyone. He points out that a few skeptics dispute the notion that Sublett actually had a gold mine in the Guadalupe Mountains, claiming that the "weathered sedimentary structure of what was once an undersea algae reef is not conducive to the formation of gold ore." Gold, according to Jameson, is formed when "hydrothermal solutions, under pressure, penetrate the rock surrounding an underground pocket of molten material." Some so-called experts claim no such activity has occurred in relation to the Guadalupe Mountains.

Jameson disagrees. Within just a few miles of the Guadalupe Mountains in almost every direction one encounters mountains and exposed landforms that are intrusively and extrusively volcanic in origin. To claim that no such activity may have taken place just below or even within the Guadalupes is folly. Furthermore, during a 1987 expedition led by Jameson into the mountains, samples of igneous intrusive rock were found on the southeast-facing slope of the escarpment, the approximate area where many believe the Lost Sublett Mine is located.

The Lost Treasure Cave: Gold and Skeletons

Somewhere along the southeast-facing slope of the Guadalupe Mountains not far from Juniper Spring is a small cave, one that has perplexed searchers for over seven decades. From the mouth of this small opening, one can see trucks and automobiles moving along U.S. Highway 62-180 approximately two miles away and at the bottom of the slope. Inside the cave lie several chests filled with gold along with a number of rifles well over a hundred years old. Sitting next to this fabulous treasure are three skeletons. How the treasure came to be concealed in this small cave, as well as the identity of the three skeletons, remains a mystery to this day.

During the early 1900s, a rancher and businessman named J.C. Hunter owned thousands of acres of the Guadalupe Mountains and adjacent foothills. Seeing great agricultural potential, Hunter stocked his property with a variety of livestock including cattle, sheep, and goats. At the time, the demand for mohair was high and profits were impressive. Ultimately, Hunter made a fortune with his large herds of angora goats.

During the late 1920s, Hunter hired a young goat herder named Jesse Duran. Duran, an experienced herder, hiked from his home in Mexico to this region in search of work and was soon placed in charge of a large herd of angoras, which grazed on the southeast slope of the range in the area of Juniper Spring.

Duran was a simple, hardworking, honest young man. He had no education and could barely read and write. His possessions were few, but his life on the Hunter ranch was satisfying, and his basic needs for shelter and food were easily met. In addition to room and board, Jesse Duran was paid approximately fifteen dollars per month for his services as a goat herder.

One day during the early part of April in 1930, Duran accidentally stumbled upon an incredible cache of treasure hidden in a shallow cave not far from the spring. The discovery was to change his life. Duran's amazing find also served to set in motion a search for this treasure, a search that continues today.

The early morning was cool and misty, a slight drizzle soaking the mountainside and a slight breeze bringing a chill to the goat herder. Under the sparse cover of a Texas madrone tree, Jesse Duran ate a cold tortilla for breakfast as he watched his herd of goats grazing on the grass that grew along the top of Rader Ridge, a somewhat narrow limestone spur that jutted for some two to three miles out from the southeast-facing slope of the Guadalupe Mountains. On finishing his breakfast, Duran noted that his canteen was empty, so he decided to hike over to Juniper Spring to fill it.

Juniper Spring was located approximately one mile from Duran's position near Rader Ridge. Confident he could leave the goats unattended for a while, he set out down a narrow trail winding across the bare limestone rock that made up these mountains. Halfway to the spring, Duran decided to take a short cut. He left the trail and worked his way down a series of sloping rock ledges, jumping from one detached limestone slab to another.

When he landed on one particular slab, his weight caused it to slide out from under him, spilling him onto the ground. As Duran stood and brushed the dirt from his pants, he was surprised to see a small opening in the limestone, one that had apparently been covered by the rock he had dislodged.

The Guadalupe Mountains, an ancient and massive limestone reef extending for almost 200 miles, is filled with caverns of all sizes, and parts of the range are honeycombed with tiny caves such as the one just encountered by Jesse Duran. The range also contains caverns of immense size, including the famous Carlsbad Caverns and Lechuguilla Cave, which may turn out to be the largest cavern network in the entire world.

Curious, Duran crouched low and peered into the interior of the dark hole he uncovered. As his eyes adjusted to the darkness, he discerned the little cave was filled with objects. He crawled forward cautiously, pausing just inside the opening for a closer look. Once inside, Duran's flesh prickled and the hair on the back of his neck rose as he recoiled in horror at what lay before him. Propped up against one of the walls of the cave were three skeletons with rotted clothes still clinging loosely to the bones and boots still on their feet. Along another wall was a stack of rifles. On the floor at the far end of the cave were several chests similar to the strongboxes used for shipment by the major stagecoach companies. The lid of one of the chests was lying open, and in the dim light of the interior of the small cave, Duran saw that it was filled with gold and silver coins.

Frightened beyond anything he had ever experienced before in his life, Duran scrambled out of the cave. Breathless, he dragged the limestone slab back into position such that it covered the opening. Shaking with fear, he continued his short journey to the spring where he filled his canteen and pondered what to do about his discovery.

Later that day Duran decided what needed to be done: He determined to report the incident to his foreman, Frank Stogden. Sometime after dark Duran arrived at Stogden's residence and requested an audience. The foreman was entertaining company—three neighboring ranchers—and they all listened intently as Duran related what happened to him near Juniper Spring that morning. When he finished, Stogden's three companions wanted the young Mexican to lead them to the treasure immediately, but Duran demurred. He told the men that, according to the Mexican tradition of *El Patron*, it is believed that the spirits of the dead occupy the space where lost treasure is found. He stated his belief that the treasure should remain where it was, and that anyone who disturbed it would likely suffer bad luck, even death.

For several minutes the men questioned Duran about the location of the treasure cave. Finally it was decided they would ride out to it in the morning, led by the young Mexican. As it was growing quite late, Stogden suggested they all turn in for the night and get an early start in the morning. Duran was invited to sleep in the barn.

The next morning as the men were saddling their horses and packing a day's supply of food, Stogden announced that Duran was gone, apparently having slipped away during the night.

Jesse Duran, frightened by what he had seen and frightened by what he believed might be severe consequences should he have a role in disturbing the treasure in the cave, fled from the area and was never seen in the Guadalupe Mountains region again.

Undeterred, Stogden and his friends set out for Juniper Spring. The men laughed and talked as they rode, discussing what they would do with their share of the fortune.

Arriving at Juniper Spring, they tied off their horses and undertook a foot search for the small cave. One of the men suggested they simply look for Duran's tracks, but the rain of the previous day had washed all sign of his passing away.

The four men searched the entire day, combing the area located between Juniper Spring and Rader Ridge. They encountered dozens of limestone slabs, any one of which could conceal the opening to a small cave, but when they turned them over or shoved them aside, they were disappointed.

The four men returned to search the next day. And the next. And the next. Off and on, they searched the area between the spring and the ridge for four months, but were they never able to find the cave.

The night he left the Stogden barn, Jesse Duran fled on foot to the home of a sister in Carlsbad, New Mexico, some 75 miles away. When he arrived, he related his recent experiences and confessed his fear of *El Patron.* Duran remained in hiding at his sister's residence for approximately four months, seldom leaving the house for fear of being spotted by Stogden or his friends. Because he feared that the spirit of *El Patron* might wreak vengeance on his family if he remained, Duran determined to leave the area altogether and traveled to California. There, he worked in the fields and orchards, married and raised a family, and passed away in 1971. Jesse Duran never returned to the Guadalupe Mountains of West Texas.

While Frank Stogden and his companions never located Jesse Duran's treasure cave, there is evidence that others have, although they never realized what they found.

Longtime Guadalupe Mountains residents Sam Hughes and Noel Kincaid, along with a few friends, were deer hunting along the southeast-facing slope of the Guadalupe Mountains one autumn. They made camp the first evening at Juniper Spring, and when dawn broke the next day, the men fanned out across the slope and foothills in search of game.

As Hughes hiked northwest of the spring toward Rader Ridge, he crossed a rugged outcrop of limestone. As he made his way across the outcrop, he weaved among numerous large limestone slabs that had been dislodged from the sedimentary layers over the centuries. Spotting a deer upslope from his position, he raised his rifle and, while trying to get a better sight, took several steps forward and fell into a small cave. Concerned that such caves might be denning places for rattlesnakes, Hughes quickly scrambled out and continued with his hunt.

Later that evening around camp as the hunters were relating their various experiences of the day, Hughes told of stumbling into the small cave opening. Kincaid, at the time the foreman of the vast Hunter Ranch empire and occasional searcher for the lost Jesse Duran treasure, asked Hughes to describe the cave and its location. Hughes' description matched that of Jesse Duran! Hughes said the cave was about a quarter of a mile northeast of Juniper Spring.

The following morning Hughes, Kincaid, and the other hunters returned to the area and searched for several hours but could never find the cave.

Treasure hunter and author W.C. Jameson writes of an encounter with yet another individual who may have found the Duran treasure, a sometime Guadalupe Mountains resident named Lester White. During the 1950s and 1960s, White, in the manner of old-time prospectors, lived in abandoned cabins and houses in the foothills of the Guadalupes. When he was evicted from these, he simply lived in his old, rusted pickup truck. From time to time White searched for different lost mines and buried treasures believed to be associated with the Guadalupe Mountains, but he had never heard of the Jesse Duran treasure until he met Jameson.

One afternoon, while drinking coffee over Jameson's campfire, White talked about finding a small cave several months earlier about one mile northeast of the Frijole ranch house and not far from a spring. At the time, the Frijole house served as headquarters for the Hunter Ranch. White said the cave was close to an old goat trail, and when he found it, it was partially covered with a large limestone slab. Curious, White tried to remove the slab so he could investigate the small cave, but being a rather frail and slight man, he was unable to budge it. Instead, he contented himself with peering into what little opening was accessible due to the position of the slab. The opening was just large enough for him to insert his head. As his eyes adjusted to the dark interior, White claimed he saw what appeared to be at least two skeletons propped up against one wall of the cave. White also noted that, hanging from the bones were old rotted clothes, and that the feet were still encased in leather boots. White had encountered skeletons several times previous in some of the many caves in the Guadalupe Mountains and was not inclined to examine these any closer, so he simply backed out and left.

After hearing White's tale, Jameson told the old man about the Jesse Duran treasure. When he finished, he also suggested to White that the Duran treasure cave and the one found by White were one and the same. Jameson asked White if he could relocate the cave, and White assured him he could.

The next morning Jameson and White set out for the location of the small cave. Once in the area, however, White became confused and disoriented, claiming that the slope looked different in the early morning light than in the afternoon light. Several small caves were spotted and investigated, but none of them turned out to be the treasure cave.

In the end, White blamed his confusion on the "angles and shadows" and concluded that one had to be in a certain location on the slope at a particular time of day in order see the opening to the cave.

As far as anyone knows, the treasure, along with the skeletons, are still in the small cave along the slope of the Guadalupe Mountains between Juniper Spring and Rader Ridge. Perhaps someday someone will be in the right location at the right time to find them.

The Big Bend

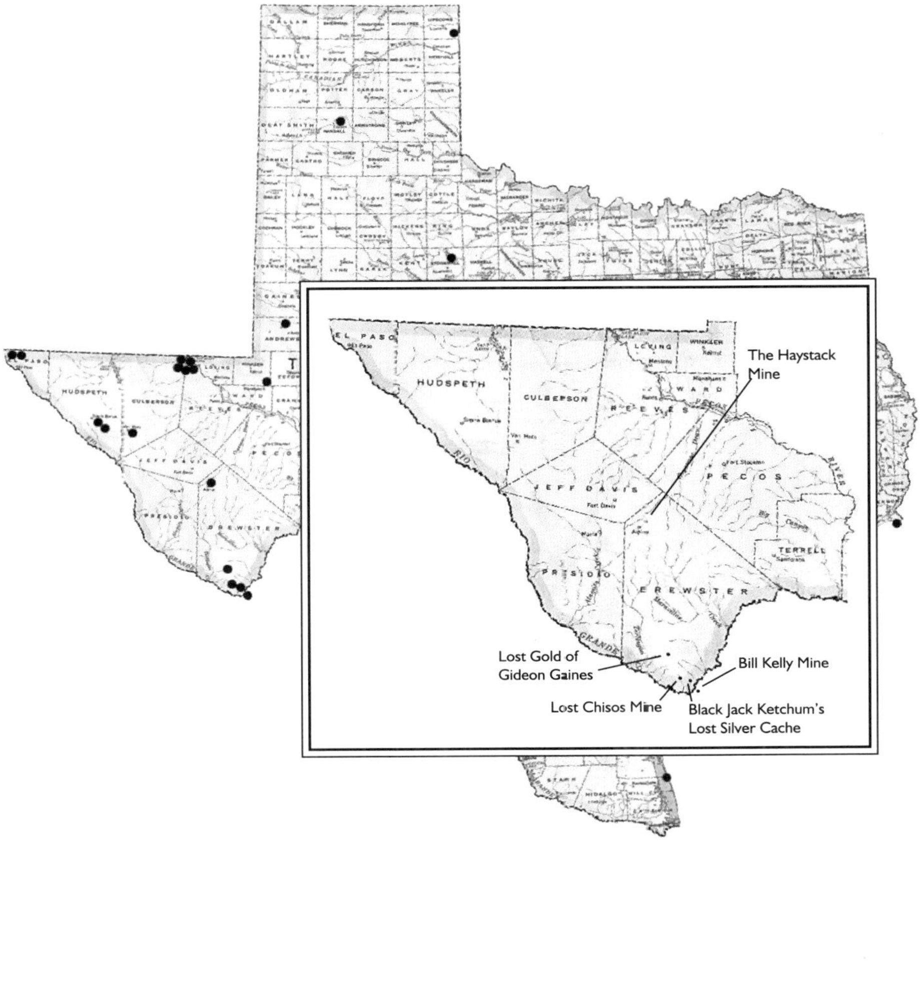
EL PASO
HUDSPETH
CULBERSON
LOVING
WINKLER
WARD
REEVES
JEFF DAVIS
PECOS
TERRELL
PRESIDIO
BREWSTER
The Haystack Mine
Lost Gold of Gideon Gaines
Bill Kelly Mine
Lost Chisos Mine
Black Jack Ketchum's Lost Silver Cache

At the southern tip of Brewster County in southwestern Texas, the Rio Grande forms a great bend in its course as it sweeps from a southeasterly direction to a northeasterly one. Brewster County is one of the largest counties in the United States and one of the least populated. Long a remote, desiccated, forbidding environment, it served as home to rattlesnakes, scorpions, mountain lions, tarantulas, sharp-pointed cacti and other desert vegetation, and tiny settlements of hardy but poor Indians who foraged and fished along the banks of the river. It was once said of the Big Bend region of Texas that everything that lived there either bit, stung, stuck, clawed, or scratched. The fearless and warlike Comanches, sometimes referred to as the "Cossacks of the Plains," often passed through the Big Bend area, rarely remaining long enough to establish permanent settlements.

The lack of water inhibited permanent Anglo and Spanish settlement here for centuries. Even today permanent settlers are few and far between.

The Big Bend of Texas has long been perceived as a rather mysterious region. Because of its remoteness and isolation it was rarely visited but often conjectured about, and it ultimately gave rise to a multitude of tales, many of which concerned lost mines and buried treasures. Because such tales are often the most enduring type to come from our culture, many associate the Big Bend of Texas entirely with such stories. The region, for all its forbidding nature, continues to lure fortune hunters into its environs—men and women who have heard the tales, are entranced by them, and are determined to search for and find these lost treasures.

Indeed, gold and silver ore, along with buried treasures, have been found in this captivating geographic area, but much remains. Following are some of the stories.

The Lost Chisos Mine

One of the greatest mysteries ever linked to the Big Bend region of West Texas is the one associated with the Lost Chisos Mine. Documented history, as well as local lore and legend, provide sufficient evidence that this mine actually existed, but for years its location has continued to elude searchers and researchers.

Some of the legends tell that gold was mined from the Chisos Mine. Others suggest it was silver. There is even one version that maintains it was both gold and silver. Whatever was taken from the mine, it is known for certain that it was operated by the Spaniards, who enslaved Indians and forced them to work in the deep, dark shafts. It is also believed that, in today's money, millions of dollars' worth of ore was removed from the mine. What is perhaps more intriguing is that it is believed by virtually all who have studied this tale that millions more remain hidden deep inside this lost mine.

During the 1770s, a time of Spanish reign throughout what is now the border region of Texas and Mexico, a number of outposts, or presidios, were established. These walled and often heavily armed fortresses represented a line of defense against marauding Indian tribes, mostly Comanches and Apaches, who ranged along the Rio Grande in West Texas. The presidios generally accommodated small settlements of Spanish, some Mexicans, and a few friendly Indians. Within the square or rectangular walls could often be found a number of rock or adobe residences, perhaps a store, and buildings within which official business was conducted. The largest building in the presidio was always the Catholic church, which was operated by one or more priests who maintained almost complete control over the local Indians and were quite influential with the Spanish military.

One such presidio was established on the southern bank of the Rio Grande almost twenty miles southeast of Emory Peak in the Chisos Mountain Range and downstream from Mariscal and San Vicente Canyons. The little settlement was named Presidio de San Vicente, and it accommodated a small contingent of soldiers.

Life in the presidio was, for the most part, boring. The soldiers rarely encountered Indians since the site of the fort was not one that was frequented by the tribes. The Comanche War Trail was located about two hours' horseback ride upriver, a trail this fierce and feared tribe followed on its frequent raids into Mexico. The Comanches seldom ventured as far downriver as the location of the presidio, thus the officers and soldiers were rarely placed in a state of alertness. They slept often, gambled, and occasionally remained drunk for days at a time, grateful for the lack of action. Enlisted men were rarely subject to drills and formation.

Settlers and soldiers alike planted corn, beans, and squash along the floodplain of the Rio Grande, and herds of goats wandered among the adjacent brush.

During such idle times, some of the soldiers would occasionally travel to the nearby Chisos Mountains, the impressive volcanic range looming on the northern horizon. In the mountains they would hunt for deer, elk, and bear to supplement the meals of goat and corn. Now and again they would kill a mountain lion for sport. During one of these trips into the mountain range, according to the legend, the soldiers found an outcrop of gold, or silver, or perhaps both in Juniper Canyon, one of several large canyons associated with the range.

After several forays into the canyon, all of which yielded an impressive quantity of ore, the presidio officers, on their own and without official sanction from the state or church, decided to commit the manpower and resources of the soldiers to the operation of the mine.

In need of a great deal of labor to enlarge and deepen the shaft, the Spanish simply made slaves of the presidio Indians. Another version of the tale claims the enslaved Indians were actually Apaches who had been captured during earlier skirmishes. Whatever the truth, the Indians were manacled together and, after being blindfolded, marched into Juniper Canyon and forced to work long, endless days digging the ore. After being led across the rickety wooden footbridge that spanned the river, the Indians, barefooted, were taken to the canyon where they remained chained together for weeks at a time. Poor food was occasionally provided, sometimes water was scarce, and the Indians were often whipped. It is said that many of them were worked to death, and that whoever finally gains access to the mine will find the skeletons of those who were left inside.

One of the priests who served at the church in the presidio noted that if one stood in the doorway of the chapel on Easter morning as the sun rose over the Sierra del Carmen range to the west on Easter morning, one would notice that the early rays would shine directly onto the entrance of the mine. To this day, searchers for what has come to be called the Lost Chisos Mine will camp at the remains of the old presidio on Easter eve in order to be up with the rising sun so they may follow the progression of the sunlight down into the canyon. One significant problem prevails, however, for the original location of the chapel is unknown.

During this time when the Spaniards were occupied with gleaning the riches from what they called the Chisos Mine, Indian raids along the border intensified, and the lack of action from the soldiers at Presidio de San Vicente was eventually brought to the notice of Spanish officials in Coahuila, Mexico. At the time, the officials were unaware of the mining activities of the San Vicente soldiers. By the time the Spanish decided to do something about the presidio, it was too late.

With the help of kinsmen who managed to evade being enslaved by the Spaniards, the Indians finally revolted, attacking their captors and killing them all without mercy. One version of the tale has the Spanish being attacked just outside the mine shaft as they were readying twenty burro loads of ore for shipment to the interior of Mexico. Another version claims the Indians then led the burros, still loaded, into the mine, slew them all, and closed the entrance to the shaft, making it look like much of the rest of the canyon. Another version suggests the Indians closed the shaft by creating a rockslide on the slopes just above it.

In a rage fueled by the killings of the Spanish miners, the Indians ran the twenty miles to Presidio de San Vicente, slaughtered all of the remaining Spaniards, burned the fortress, and then left, never to return. When a contingent of Spanish soldiers passed through this area in 1781, all they found were the charred remains of the settlement. Human bones littered the ground inside the fortress. Human skulls, apparently decapitated from the bodies, had been placed atop fence posts and adobe walls.

Though the Spanish and the Indians are now long gone from the area, the specter of the Lost Chisos Mine remains. Within years after the Spaniards were killed or driven from the region, the story of this lost mine began to spread throughout this part of the Southwest. Early settlers to this region, and they were few, often hunted for the location of the mine when they were not busy trying to eke out a living on the arid expanses of the Chisos Mountain foothills and the surrounding plains of the Chihuahuan Desert.

Many years ago a Big Bend rancher found an old gentleman who claimed to be a former resident of the Presidio de San Vicente. The old man told the rancher that his grandfather told him that if you stood among the ruins of the old fortress and looked toward the Chisos Mountains to the northwest, the mine lies in an area where the rock appears the whitest. Indeed, as a result of the weathering of the granite that makes up a large portion of this part of the range, patches of rock that appear to be white in color are found here and there. A number of researchers are convinced that the Lost Chisos Mine one of the canyons that lead up to a granite prominence that has been named Lost Mine Peak.

To add to the likelihood that the lost mine may be in this area, in 1914 another Big Bend rancher, a man named W. T. Burnham, found an old bar of silver near Lost Mine Peak. The ingot, according to Burnham, bore Spanish markings!

In the early 1900s this region was rigorously surveyed by a mining company. During the survey, several traces of silver were found. Furthermore, evidence of an old trail leading from the Presidio de San Vicente and winding across the desert floor up into Juniper Spring Canyon is visible from the air and even from a few locations on the ground. At its upper reaches, the trail simply ends, its terminus apparently covered by a rockslide. In this area, another Big Bend rancher named Lloyd Wade found yet another ingot of silver.

Given the available evidence, it appears very likely that the Lost Chisos Mine did, in fact, exist, and that it was run by the Spaniards using Indian slave labor. Additionally, given the evidence of the silver ingots found in the vicinity, it would be easy to conclude that the mine yielded silver ore. While the reported rockslide that covered the mine shaft occurred in excess of two centuries ago, several professional geologists and geographers who have examined the site concur that enough evidence exists to suggest that at least one, and perhaps several, rockslides did occur.

Given the facts, and given the enduring lore that has prevailed in this area for generations, it is easy to conclude that the so-called Lost Chisos Mine did exist. It likely exists today beneath the rockslide debris in the higher reaches of Juniper Spring Canyon, perhaps still containing a wealth of silver as well as the skeletons of the Indians who labored and died in the mine.

The Lost Gold of Gideon Gaines

That gold can be found in the Chisos Mountains of the Big Bend area of Texas is certainly evidenced by the tale of Gideon Gaines.

Gaines, along with a partner named Hurley, discovered a vein of gold sometime during the 1820s or 1830s in the Chisos Mountains. Based on what old-timers remembered when interviewed, the consensus is that the ore was found at some long lost location on the northeastern side of the range. Though never authenticated, there exist a number of second- and third-hand tales relating to the notion that several people had seen the gold that Gaines and Hurley removed from their mine.

Near this lost mine, Gaines constructed a small arrastre to aid in separating the ore from the quartz matrix. Once cleaned from the quartz, the ore was then melted and poured into crudely made molds to form ingots. On at least two occasions, Gaines and his companion transported the ingots into Fort Stockton, several days' ride to the northeast, where they were converted into cash. The gold was reputed to be quite pure, and comments made by Gaines during the times of these transactions led others to believe that the mine was very rich and apparently held a nearly inexhaustible source of the ore.

During this early part of the nineteenth century, Apache and Comanche Indians roamed freely in and around the Chisos Mountains. While the Comanches normally passed through on their way into Mexico from time to time, the Apaches actually lived in the mountains for months at a time. The Apaches were suspected of conducting raids on nearby settlements, and many believed these Indians were responsible for the killings of several travelers, trappers, and prospectors.

Gaines and his partner often expressed concern over the presence of the Indians, but the lure of wealth from the gold was greater than their fear of death at the hands of others. The two miners occasionally spotted Indians watching them from a nearby ridge but for the most part were not bothered by them.

On occasion Gaines reported spotting Indians watching his camp from vantage points high on the ridge, and when he and his partner traveled to Fort Stockton or Alpine, they were sometimes followed by two or three of the warriors.

One morning, just after Gaines and his partner loaded three burros with packs filled with ingots, he noticed that the numbers of Indians watching from the ridge had doubled. He considered this a bad omen, and as he and his partner led the heavily laden burros down the long, winding trail to the flats, they noticed the Indians, on horseback, were following them, remaining at a distance of approximately 100 yards.

When Gaines was still about one-quarter of a mile from the bottom of the slope, he turned in the saddle and was shocked to discover the Apaches were rapidly closing the distance between them and were readying their weapons. He warned his partner that the Indians were preparing to attack. Mistakenly believing that the Apaches were after the gold they carried, Gaines decided that, while they were briefly hidden by an upcoming turn in the trail, they would quickly hide the ore, turn the burros loose, and ride away as fast as they could.

Just around the turn and partially concealed behind a jumble of rock was a spring, one from which Gaines had filled canteens and jugs in the past. The spring had a wide opening; the pool of water that formed at the top was about three feet in diameter. Riding quickly to the spring and jumping from their horses, Gaines and Hurley quickly cut the straps that secured the gold-laden packs to the burros and one by one threw them into the spring, watching them sink out of sight. Leaving the burros, the two men remounted their horses and rode away down the trail as fast as they could. Their horses, unfortunately, were not the sturdy, mountain-bred mounts such as those ridden by the Apaches, and before two minutes had elapsed, the Indians were upon them. Almost immediately Hurley was struck by several arrows and fell from his horse mortally wounded. Terrified, Gaines fired his revolver at the Indians until he ran out of bullets. At this close range, he was certain he hit at least two of his attackers. Gaines decided to find some cover from which he could defend himself. As he was dismounting, he took an arrow in the thigh. In great pain, he scrambled toward the shelter of some nearby rocks.

Gideon Gaines never made it. As he alternately ran and crawled along the slope, the Apaches overtook him and clubbed him until he lost consciousness.

Moments later as his head began clearing a bit, he felt the Indians pulling off his boots and clothes. He pretended he was dead in the hope that the Apaches would be satisfied and leave. As he lay naked and still on the ground, Gaines heard the Indians likewise fussing over Hurley's body.

A while later one of the Indians returned to where Gaines lay, grabbed a handful of his hair, pulled his head up, and scalped him. The pain of the scalping was so intense that Gaines could barely keep from screaming, but he knew that to do so would mean death for certain. After his scalp was ripped away, the Indian dropped Gaines back to the ground and walked away. A few minutes later Gaines could hear the Indians riding away down the trail.

Gaines waited motionless for another ten minutes before daring to move. Presently he opened his eyes, looked around, and convinced he was alone, tried to rise. His body was covered with blood from his bludgeoning, from two arrow wounds, and from the scalping. Unsteadily, he rose to his feet, and the first thing he saw was the naked, scalped, and mutilated body of his friend Hurley. Hurley's death, along with his own pain and loss of blood was too much for Gaines, and he collapsed to the ground in a faint.

Gaines did not awake until the following morning. Forcing himself once again to his feet, he realized he was suffering an intense thirst. He decided to try to make his way back up to the spring where they hid the gold. There, he would slake his thirst, wash the blood that was caked on much of his body, and make plans to somehow reach civilization. As he began hiking up the trail, he noticed a leather pouch the Indians somehow missed. Picking it up, he realized it was Hurley's. Inside were a variety of things, including a can of tobacco, a lead bullet mold, and some scraps of paper. Dragging the pouch, Gaines began crawling his way back up the trail toward the spring.

After several yards, Gaines was forced to stop and rest. While seated in the shade of a juniper, he spied the body of a dead Indian, likely one of the ones he shot. With great difficulty, Gaines removed the filthy, blood-soaked cloth pants worn by the Indian and put them on himself. Under a second dead Indian he found nearby, he discovered a knife, which he appropriated.

A few more paces up the trail, Gaines came across a dead burro. Using the knife he found, he sliced some meat from the animal's rump and cooked it over a low fire he managed to build. Refreshed somewhat from the meal, Gaines resumed his quest to reach the spring.

Gaines never arrived at the spring, and evidence encountered by two travelers several years later indicated he simply perished on the way. While riding along a trail in the Chisos Mountains, two men named Flippin and Breede came upon a human skeleton. As they were examining the bones, one of the men found a metal tobacco can nearby, inside of which was a scrawled note on several pieces of paper. The note detailed the attack by the Apaches, the scalping, and a valiant attempt to survive by the writer. It also told of throwing the packs of gold into the spring and that the writer of the note was trying to return to it. The note was signed, "Gideon Gaines."

Flippin and Breede spent the better part of a week trying to find the spring mentioned in the note but were unsuccessful. During the ensuing years, they told their story to others, and men came to the area in search of what has come to be called the Lost Gold of Gideon Gaines.

While all available information suggests Gaines and Hurley did indeed find gold in impressive quantities in the mountains, the exact location of the spring remains a mystery.

Springs often dry up due to drought, minor shifts in the earth's crust, and numerous other reasons. Sometimes they are covered over by rockslides. Quite often when a spring dries up there generally remains adjacent to it telltale evidence that one was once there, evidence in the form of vegetation growth greater than in the surrounding area. This is especially noticeable in deserts. However, with the passage of nearly two centuries, it is entirely possible that even that kind of evidence may have long since disappeared.

Whoever may be fortunate enough to stumble onto Gaines' lost gold will no doubt become a wealthy individual, for the fortune in gold is likely still there. Furthermore, the mine from which the gold came is also there, somewhere farther up that same trail, luring and tempting searchers and waiting to be rediscovered.

The Curse of the Lost Bill Kelly Mine

Bill Kelly was as unlikely a person to discover a huge deposit of gold as anyone. The discovery changed his life, as well as the lives of several of those around him.

Kelly was a half-breed, a Seminole-Negro, who had wandered into the Big Bend area of Texas from his home far to the south in the Mexican state of Coahuila. Dressed in rags and barely able to speak English, Kelly nevertheless secured a job as a cowhand on a Big Bend area ranch. He proved to be a good worker, an excellent horseman, a dependable hand, and an honest man.

Kelly was hanging around the Southern Pacific Railroad loading pens in Dryden, Texas, one day when the Reagan brothers—Frank, Jim, John, and Lee—arrived with a large herd of cattle. They had just completed the trek from their ranch some seventy-five miles to the southwest. As they were penning the cattle, Kelly, only nineteen years old at the time, approached the men and inquired about work. They agreed to let him help them load the cattle, and before the day was over they became impressed with his ability to handle livestock. They offered him a full-time job on their ranch, and after providing him with a horse, the Reagan brothers and Kelly rode back to the ranch the following morning.

Kelly proved to be one of the best hands the Reagan brothers ever hired. He rose with the sun, worked hard all day tending cattle and horses, mending fence, and performing other chores, and only returned late at night when it was too dark to see.

Late one evening as Jim Reagan handed Kelly a plate of food on the back porch of the ranch house, the black man mentioned that, while conducting an unsuccessful search for some horses that strayed across the river and over into Mexico, he found what he believed was a gold mine. Reagan, certain Kelly wouldn't know anything about gold, simply ignored the statement and reentered the house.

The following morning Kelly rode back across the border to look for the lost horses, this time accompanied by Lee Reagan. As they paused to rest in the shade of a canyon wall late in the day, Kelly commented to Reagan that the mine he found the previous day was only a short distance from where they sat. Lee, like his brother Jim, could not believe that Kelly would know gold if he saw it, and replied that he thought they ought to be getting back to the ranch. At that, Kelly reached into his saddlebag and withdrew a piece of gold ore. Later Lee Reagan described it as being "as big as the eye of a horse." Kelly handed it to Reagan, but the rancher, at the time unconvinced that it was gold, simply tossed it onto the ground and told Kelly they were wasting time.

About one month later the Reagans, along with Kelly, drove another herd of cattle to the loading pens at Dryden. When the job was completed, Kelly asked for permission to travel to San Antonio to visit his brother for a few days. The Reagans agreed, and when the train departed, Kelly was on it.

While seated in a passenger car, Kelly entered into a conversation with the conductor, a man named Campbell. He related his discovery of gold on the Mexican side of the Rio Grande and gave Campbell general directions for its location. Once in San Antonio, Kelly handed Campbell a piece of quartz thickly laced with gold he claimed came from the mine. Campbell told Kelly he would have it analyzed and let him know what he found out. While in San Antonio, Kelly left another piece of ore with an assayer named Fisher.

Two weeks later Kelly returned to the Reagan ranch and went back to work. Several more weeks passed, and while picking up the mail in town, one of the Reagan brothers noticed a letter addressed to Kelly. Curious as to why anyone would write to an illiterate half-breed ranch hand, the brother opened the letter and read the report from the assayer Fisher. Fisher had written to Kelly to inform him that his gold sample was impressively rich and that it assayed out to $80,000 per ton. The brother folded the letter back up, placed it in his pocket, and hurried back to the ranch to tell his brothers the news.

Two days later when Kelly rode in from checking cattle in a distant canyon, the Reagans' cook told him that the brothers opened and read his letter and were plotting to get him to lead them to the mine. The cook told the black man he was concerned that the men intended to kill him after the mine was located.

That evening after everyone went to bed, Kelly slipped into the corral, saddled a horse and rode away into Mexico. The following afternoon he arrived at the Coahuila ranch of a man named Chessman and asked for a job. As he was with the Reagan brothers, Kelly proved to be an excellent worker. Chessman ranch foreman John Stillwell was especially impressed with Kelly, and the two men eventually became good friends.

One evening Kelly told Stillwell about the gold mine and about the plot hatched by the Reagan brothers to locate it. Kelly confessed he was worried the Reagans would try to find him and kill him. When Stillwell expressed doubts about the existence of a gold mine, he later related, Kelly walked over to his saddlebags and retrieved a handful of rich gold nuggets he claimed came from the location he spoke of.

About two months later Kelly accompanied Stillwell and several other cowboys on a cattle drive to Mexico City. After the cattle were delivered, Bill Kelly vanished.

Meanwhile, Campbell, the train conductor who earlier befriended Kelly, learned that the gold given to him by the cowhand was extremely pure and rich. He made several attempts to contact Kelly and eventually learned that the ranch hand had ridden away from his job at the Chessman ranch and was never heard from again. Some five years later Campbell encountered Jim Reagan at a cattleman's conference in San Antonio and told him about the gold mine. Reagan convinced Campbell he was certain he could locate the mine. Weeks later Campbell traveled to the Reagan ranch and, with the brothers, undertook a search for the mine. Lee Reagan led Campbell to the place where Kelly had once told him they were not far from the mine, but though they searched for several days, nothing was found. Over the next five years, the Reagan brothers continued to search for the mine, occasionally accompanied by Campbell, but its location continued to elude them.

After the Reagans had all but given up hope of finding the gold mine, a black man walked into a mercantile in Eagle Pass, Texas, dumped a pouch of gold nuggets on the counter, and said he would exchange the ore for $1,000. The owner of the store examined the gold, found it to be very pure, and determined it was worth far more than $1,000. When he asked where the gold came from, the black man claimed it came from an old mine located across the Rio Grande in a range called the Ladrones Mountains. The storekeeper said he would agree to the deal only if the black man would lead him to the source of the gold. The man refused and walked out of the store. Later the proprietor learned that the man was Bill Kelly.

During the summer of 1899, the Reagan brothers partnered with Campbell, fellow ranchers D.C. Bourland, L.C. Mueller, and an experienced prospector and miner named John Finky. Together, the men planned a systematic search in the Ladrones Mountains for the source of Bill Kelly's gold. Finky was supplied with provisions, a place to live, and a small stipend while he looked for the mine.

One afternoon Finky returned from his search and informed Bourland that he had found the mine. He told them that, while exploring a remote canyon in the Ladrones Mountains, he came upon a skeleton. Next to the skeleton was a leather pouch containing a number of large gold nuggets. Intrigued, Finky continued to explore up the canyon, eventually topping out on a low ridge where he found the mine. The gold was obvious and plentiful, he said, and proved it by dumping out a saddlebag full of the ore for Bourland's inspection.

Finky suggested the other partners be contacted and a return trip to the mine undertaken. Bourland told Finky that the trip would have to wait for a few weeks—at the time, that part of West Texas was undergoing a severe drought, and the ranchers were kept busy night and day trying to provide for their livestock. Finky saw the logic in this and agreed to wait.

That same evening as Finky lay on his cot, he was stung by a scorpion. The following morning his face and neck were grotesquely swollen, and Bourland transported him to a physician several miles away in Sanderson. Finky was treated and the doctor recommended he remain hospitalized for a few days for observation.

The next day Finky was visited in the hospital by Bourland and Jim Reagan. The two ranchers tried to get Finky to give the directions to the mine, but he refused. Finky did not trust the men and feared they planned to exclude him from the partnership once they learned the location of the gold mine.

Finky also informed the men that they could not legally conduct a mining operation in Mexico without permission from that country's government. He told them that when he recovered, he would travel to Mexico City himself to obtain the necessary permits.

It was two weeks before Finky finally recovered from the scorpion sting, and when he was released from the hospital, he journeyed to El Paso where he intended to catch a train to Mexico City. On his first night in El Paso, Finky went to a tavern and began drinking heavily. Later that night he mentioned to the bartender that he was suffering from an excruciating headache. The bartender arranged for Finky to rent a room for the night in the establishment. After a few more drinks, Finky went to his room to sleep. In the morning they found him dead.

Sometime in the early 1900s, a mysterious stranger was often seen exploring around the Chisos Mountains of the Big Bend. The few people he spoke with learned that he possessed a map that purported to show the location of a rich gold mine located somewhere in the Ladrones Mountains. It was further learned that the stranger received the map from a relative who was on death row at an Oklahoma prison.

The relative, according to the tale that was subsequently gleaned from the stranger, had been involved in stealing horses in Mexico and selling them in Texas. During one such horse-stealing trip, the relative and three companions were herding stolen horses through the maze of canyons in the Ladrones Mountains. Hot on their trail was a contingent of Mexican soldiers intent on capturing and killing the horse thieves. In one wide canyon, the riders encountered an old man transporting two heavy leather packs on his back. They stopped the old man to rob him, opened the packs, and were surprised to find both filled with gold ore. At gunpoint, they forced the old man to show them the location of the mine from which the gold had come. In spite of the posse behind them, the horse thieves followed the old man up a nearby canyon and thence to a low ridge. There, in plain sight, was the mine. Next to a pile of scree was a vertical shaft that extended several feet into the solid bedrock. In the light of the afternoon sun, the horse thieves could peer into the shaft and see the reflection off the thick veins of gold that wove through the quartz matrix. After taking note of surrounding landmarks to facilitate their return to the mine, the horse thieves shot the old man and threw his body over the rim and into the canyon. It is likely that this was the skeleton found by John Finky.

During his search for the gold mine, the stranger met and was befriended by area resident John Young. On seeing the map, Young, who was quite familiar with the Ladrones Mountains, verified that all of the landmarks denoted thereon were correct. Young, the stranger, and a third man named Felix Lowe formed a partnership to search for the mine. Young traveled immediately to Mexico City to secure permission to conduct a mining operation in the Ladrones Mountains.

After eventually receiving the appropriate authority, the three men, leading packhorses carrying equipment and supplies, crossed the Rio Grande and entered the range. Once there, however, they became confused about the directions and began arguing among themselves. After three months of unsuccessful searching, they finally returned to Texas to make plans for another expedition.

The year was 1910, and as the stranger, Young, and Lowe were revising their plans to search for the gold mine, the Mexican Revolution broke out and the border became a dangerous place. Frustrated, the three decided to abandon the search.

At this point, what has come to be called the Lost Bill Kelly Mine was essentially unknown save to only a few individuals. Several months following the outbreak of the Mexican Revolution, Young was interviewed by a San Antonio newspaper reporter, told the story of the lost mine, and the tale became widely known in a very short time.

Sometime after the article on the mine appeared, Young received a letter from a California rancher named Jack Haggard. Haggard claimed Bill Kelly worked for him for a time and revealed the location of the mine to his foreman. The foreman went to look for the mine and apparently found it, for he returned to California with a number of large nuggets of very pure gold. The foreman told Haggard there was enough gold in the mine to make several men very rich, so he quit his job and made plans to return to the Ladrones Mountains. While working for Haggard, the foreman also worked part-time in a nearby mine. In the process of finishing out the week, however, he was killed in a freak accident.

So impressed was he by the gold nuggets he saw, Haggard traveled to the Ladrones Mountains to search for the mine. After spending nearly a month in the range, he finally found the mine only to be chased back across the river by Mexican troops. As he was making plans to return when the revolution was winding down, Haggard drowned during a fishing trip.

In 1915 Big Bend area ranch hand Will Stillwell traveled to Marathon, Texas, and told his brother, Roy, that he had found the Lost Bill Kelly Mine. While making his way through the Ladrones canyons following a cattle-buying trip deep in Mexico, Stillwell and his partner came upon an old Mexican woman who had apparently been left out in the desert to die. After giving her water and food and treating her badly blistered feet, they carried her to a small nearby Mexican settlement. They remained there for two days to make certain she would recover, and when she did, she told the two men that she wanted to thank them by providing them with the location of a rich gold mine. As she described the mine, Stillwell took notes and sketched a map. She said to go to a particular canyon that was known for its abundance of petrified tree stumps. She said deeper in the canyon and past the stumps they would encounter a very large boulder. On the left side of the boulder, she said, they would find a hatchet that had been buried only a few inches below the surface. On the opposite side of the boulder, they would find two packs filled with gold. The mine from which this gold came, she continued, was a few hundred more yards up the canyon and located on a low ridge. The mine, according to the old woman, had been filled in with rocks and dirt.

Stillwell and his partner located the canyon with the petrified tree stumps, the boulder, and the hatchet. The gold-filled packs, however, were not where the old woman said they would be. They continued up the canyon to the low ridge and walked right up to the mine, which was partially filled with debris and rocks.

The two men decided to dig the debris out of the shaft and returned down the canyon to a spring where they intended to fill their canteens. While riding toward the spring, Stillwell and his partner were suddenly surprised by several Mexicans on horseback. The Mexicans, according to Stillwell, appeared to be bandits and gave chase. While fleeing from the bandits, Stillwell and his partner rode through the maze of canyons in the Ladrones

Mountains for two days. Finally they arrived at the Rio Grande and swam their horses to safety on the opposite shore. Will later told his brother that once the Mexican Revolution was over, he intended to return to the Ladrones Mountains to dig the gold out of the old mine he had found there.

Several weeks later Stillwell joined a company of Texas Rangers and was given an assignment patrolling the border in the Big Bend area. Settlers in the region were continually plagued by bandit raids, and it was the responsibility of the rangers to protect them. While stopping at a small trading post at Castolon, Will Stillwell was shot in the back by a Mexican outlaw and died instantly.

Roy Stillwell, however, still retained the map to the Bill Kelly Mine that had been given to him by his brother. On learning of Roy's possession of the map, several investors approached him about financing an expedition to the mine. Roy refused, claiming he was concerned there might be a curse on the mine. He pointed out that all who had tried to retrieve the gold met with death.

Then one day a group of investors made Roy an attractive offer of money that he found difficult to decline. His share of the money would represent a sum that would keep him and his family comfortable for the rest of his life. Roy told the investors he was very interested but he wanted to talk it over with his wife and think about it for a week before giving his answer.

Two days later, however, Roy Stillwell was killed when a truck turned over on him. The map that was in his possession was never found.

In 1946 it was learned from a relative that when Bill Kelly fled from the Reagan brothers' ranch he returned to his hometown in Coahuila. A few years later he traveled to Oklahoma, became involved in trafficking illegal liquor, and was sent to prison for several years. Upon release, he went to California where he worked from time to time on three separate ranches. Eventually, he moved to San Antonio where he lived until he passed away, it is believed, sometime in the late 1930s or early 1940s.

In 1947 two men identifying themselves as mining engineers arrived at the Reagan ranch asking for directions to the Ladrones Mountains. After receiving directions, they spent several weeks exploring the range and then departed. The pair returned six months later with a truckload of mining equipment and made an agreement with Marathon resident and sometime cowhand Ed Shirley to help them pack into the Ladrones Mountain as well as perform critical camp chores and care for the horses and mules. While the two engineers never told Shirley directly, the cowhand overheard enough of their conversations to know they were searching for the Lost Bill Kelly Mine.

One afternoon while Shirley was cleaning up the campsite that had been established in the Ladrones Mountains, the two engineers rode in. They were very excited about something they apparently found at the top of a nearby low ridge, but they refused to tell Shirley what it was. Later that afternoon they paid Shirley off and sent him away.

Years later Shirley stated he was convinced the two mining engineers had discovered the Lost Bill Kelly Mine. For several weeks following his dismissal, Shirley hung around the Reagan ranch waiting for the two men to return from the mountains. He was curious to look upon the gold he believed they found.

They never returned, nor were they ever seen or heard from again, and what became of them remains an ongoing Big Bend mystery to this day.

Treasure hunter and author W.C. Jameson has spent a great deal of time in the Ladrones Mountains of Mexico. Even under the best of conditions, he says, the Ladrones are a rugged and forbidding place, and survival is often tenuous at best. Springs dry up, game can grow scarce, and there is the ever-present danger of rattlesnakes, mountain lions, bears, wolves, and outlaws. It is quite likely that the two engineers succumbed to the hardships of the mountain range, or they may even have suffered an accident while working in the mine.

Or, according to Jameson, they may have been the victims of the curse of the Lost Bill Kelly Mine.

The Haystack Mine

About sixty miles north of the Chisos Mountains in Brewster County, Texas, lies the town of Alpine. Alpine is a relatively quiet place nestled among the granite mountains and hills in this part of the Chihuahuan Desert. Amazingly, there exists about a half-dozen interesting tales of lost mines that supposedly lie within just a few miles of Alpine. The details relative to these mines are, for the most part, elusive and sketchy, and while the stories are quite provoking and have led to numerous searches, little of significance has ever turned up.

One notable exception exists with what has come to be called the Lost Haystack Mine, a location that may very likely be one of the richest gold deposits ever found in the state of Texas. That the mine yielded an impressive quantity of rich gold is not doubted—it has been seen by a number of people. What remains a mystery is the exact location of this incredible mine.

Sometime during the last decade of the nineteenth century, a Seminole-Negro named Mister Johns, along with his wife, migrated from his home in Coahuila across the Rio Grande into Texas where he found work at various ranches as a sheepherder.

Mister Johns built a crude shack between the young settlement of Alpine and the Haystack Mountains, which lay only a few miles northwest of the town near the grassy foothills of the mountain where he tended sheep. In time Mister Johns and his wife had several children and, eventually, grandchildren, all of whom also found gainful employment with the area sheep ranchers.

By this time Mister Johns was, it is believed, close to eighty years of age. His favorite grandchild was a boy named Jim Johns. Jim Johns loved his grandfather dearly and constantly followed the old man around as he tended his herd of sheep. From time to time during the day, Mister Johns and Jim Johns would take a break and lie in the shade of nearby trees having lunch or taking short naps.

One rainy morning Mister Johns noticed that a number of sheep were missing from the herd. Assuming they had wandered up a narrow canyon of Haystack Mountain, he mounted his horse and set off in search of them. Within minutes after entering the canyon, the storm grew fierce and the rain fell in sheets. Spotting a narrow opening in an adjacent canyon wall, Mister Johns decided to take shelter there and wait out the storm. After tying his horse to the limb of an oak tree, Mister Johns crawled into the tiny opening and out of the rain and began removing his wet garments.

As he took off his coat, Mister Johns noticed that he was in a small natural cavern and that a narrow passageway extended deeper into the mountain. Mister Johns decided to explore the cave while waiting out the rainstorm. After fashioning a crude torch from dry materials he found near the opening, he proceeded down the passageway.

When Johns had crawled about ten yards, the passageway opened up into a small room just barely large enough for him to stand up. In the light of his torch, he saw stacked along one wall of the room a number of leather sacks, the pile covered with a deep layer of dust. Kneeling before the pile, he cut open one of the sacks with his knife and watched in surprise as a handful of gold nuggets tumbled out onto the floor, the ore gleaming brightly in the flickering light of the torch. He cut open several more sacks and found that they all contained gold.

Mister Johns was alternately excited about and terrified by his discovery. It was slowly dawning on him that he was wealthy beyond his wildest dreams. On the other hand, if he showed up in town with a large amount of gold or money, he knew he would surely be accused of stealing it. He did not know what to do. Finally he decided to leave the gold where he found it for the time being while he pondered what course to take. By this time the storm had lessened, and Mister Johns returned to the cave opening, put on his coat, untied his horse, and continued searching for the lost sheep.

When he returned home that evening, Mister Johns informed his wife of his amazing discovery. Like him, she thought it best to leave the gold in the cave until some time later when it would be more appropriate to retrieve it.

Only a few months later, however, Mister Johns, who was already aged and in increasingly poor health, was stricken with a bout of pneumonia that confined him to bed. When a local physician was called in to examine him, he told Mrs. Johns that the old man only had a few more days to live.

Knowing he would die soon, Mister Johns summoned his grandson Jim Johns to his bedside and told him about finding the huge gold cache in the little cave in Haystack Mountain. Though his memory was dimmed somewhat with age and sickness, Mister Johns provided Jim Johns with vague directions to the cave.

Jim Johns, now a young man, waited a few years before trying to find the cave, for like his grandfather he was concerned about people thinking the gold was stolen. Five years from the time Mister Johns passed away, Jim Johns, after several unsuccessful searches, finally located the treasure cave. Coming upon the leather sacks filled with gold, he transferred several pounds of the ore to his own knapsack, returned to Alpine, purchased a bus ticket, and traveled to San Antonio where he converted the gold to cash.

Jim Johns knew he could never return to Alpine a wealthy man. Known by everyone there as only a poor Seminole-Negro sheepherder, he would surely arouse the suspicion of the townsfolk and likely be followed everywhere he went. Instead, he chose to remain in San Antonio and live off of his newfound wealth.

When Jim Johns exhausted his supply of money after several months in San Antonio, he simply boarded a bus, traveled back to Alpine, and returned to the treasure cave to refill his knapsack with gold. During the decade of the 1930s, Jim Johns returned to Alpine several times, often stopping to visit with friends and family, and occasionally showing them some of the gold he took from the secret cache.

During one such visit, Jim Johns pulled a handful of gold from his knapsack and showed it to some of his friends. Without providing any information relative to the location of the gold cache, he told them the story of the treasure cave that was discovered by his late grandfather many years earlier. Before long, the story of Jim Johns' gold was on the lips of most Brewster County, Texas, residents.

During one of his visits to Alpine in 1937, Jim Johns was approached by several men who offered to pay him an amount of money for information pertaining to the location of the gold cache they had heard about. Johns informed them that the gold in his knapsack represented more than ten times the amount of their offer and turned them down.

Several days later the same group of men met with Johns and made another offer, this one somewhat larger. Johns again refused, and the men grew angry. An argument broke out, and before it was over, Jim Johns had been shot in the stomach. After an operation and three weeks recovering in the hospital, Johns vowed never to return to Alpine for fear of being followed and killed. He took what gold he had and fled to San Antonio for good. One year later Jim Johns died from a heart attack. As far as is known, he never revealed to another soul the location of the treasure cave.

That the gold existed has never been questioned—on several occasions, dozens of people saw and handled the rich ore Jim Johns carried in his knapsack. Furthermore, during his visits Johns was often seen leaving Alpine and walking toward Haystack Mountain. The next day he always returned with several pounds of gold.

One mystery that has never been solved is the source of the gold—how it arrived at the small cave in Haystack Mountain remains unclear. Some believe it is Spanish gold from the time of organized explorations into this area. Others suggest it represents loot from stagecoach holdups.

Another mystery, one that has treasure hunters scratching their heads today, is the location of the treasure cave in Haystack Mountain. Dozens of Alpine residents have tried to find the cave but with no success. Professional treasure hunters have also come away from the search empty-handed.

According to the original descriptions of the gold cache provided by Mister Johns and Jim Johns, a sizeable fortune remains in this elusive and mysterious cave.

Black Jack Ketchum's Lost Silver Cache

Black Jack Ketchum, notorious western outlaw, once robbed a train in the Big Bend region of West Texas, escaped with over $6,000 in silver, and cached it in a small cave deep in the Chisos Mountains. On the run from a pursuing posse, neither Ketchum nor his henchmen were able to return to the region to retrieve the treasure.

It was approximately two o'clock in the morning of May 10, 1897, when a Southern Pacific Sunset Limited stopped at the isolated and remote Lozier Station in Val Verde County, Texas. Bound for El Paso, the train had left Del Rio, Texas, only a short time earlier.

While climbing the uphill grade out of Del Rio, the engineer, a man named Freese, told conductor Burns the engine was not running smoothly and asked him to pull to a stop at Lozier, the next station down the line. While Burns and Freese checked on the engine, the train's fireman topped off the water tank. In the express car farther down the track, Wells Fargo messenger William Henry Boyce inventoried the load that had been placed on board in Del Rio. Just over $90,000 in paper money, along with at least $6,000 in silver coins, was packed into the safe, which was then closed and locked.

While all of this activity was going on, three men lurked in the shadows just behind the station building. One of the men was Thomas "Black Jack" Ketchum. Ketchum had a well-earned reputation as an outlaw in Wyoming and Colorado, but he had drifted south following several close encounters with lawmen.

With the engine running smoothly, Burns gave the order to continue and the train eased away from the station. Just as it was picking up momentum, Ketchum and one of his companions leaped onto one of the cars, scrambled onto the roof, and carefully made their way toward the engine. As this was transpiring, the third member of the gang was cutting the telegraph wires at Lozier Station. Following this, he ran to a shallow ravine about fifty yards away, climbed aboard his mount, and rode alongside the tracks leading two horses.

Several miles west of Lozier, Ketchum swung down from the coal tender into the cab of the locomotive, pointed a rifle at the engineer, and commanded him to halt the train. Leaving his companion to guard the conductor, Ketchum forced the fireman, a man named Bochat, and the engineer, Freese, out of the cab. Dropping to the ground himself, he ordered them to lead him to the express car.

The chuffing of the idling steam engine nearly drowned out the sound of the men's boots as they crunched along the gravel in the darkness. On reaching the express car, Ketchum rapped loudly on the door with the butt of his rifle and called out for the messenger to open it, but there was no response from within. Twice more he called, and when he still received no response Ketchum crawled under the car and fired a shot up through the wooden floor. Ketchum then told Bochat to order Boyce to open the door or he would shoot the engineer. After cocking the rifle, Ketchum pointed the barrel at the engineer's chest. Bochat, worried that Freese would be harmed, pleaded with the messenger. Several seconds later the express car door slid open and, with Boyce standing on board with his hands raised, Black Jack Ketchum climbed in.

Placing the end of the rifle barrel against Boyce's neck, Ketchum ordered him to open the safe immediately. Boyce explained that he did not know the combination to the safe, and that the only person who did was the relief messenger in El Paso.

Angered, Ketchum knocked Boyce to the floor. The outlaw then reached into a deep coat pocket and pulled out a bundle of four sticks of dynamite that had been tied together earlier. After attaching the explosives to the door of the metal safe, he lit the fuse and, pulling Boyce with him, jumped out of the car. A few moments later the blast rocked the express car. When the smoke cleared, Ketchum reentered the car and found the metal door completely blown off and the contents of the safe exposed. Finding the silver coins, Ketchum immediately began stuffing them into a knapsack he produced from another coat pocket. The $90,000 in currency had been packed so that it would appear like newspapers, and Ketchum overlooked it.

Ninety minutes from the time the train had been stopped, the robbery was completed. Leaving the engineer, the conductor, the fireman, and the messenger standing by the express car in the darkness of the early morning, Black Jack Ketchum and his two unidentified companions rode away to the south.

Moments later the train was speeding to the next station up the track at Sanderson. Stopping there, the conductor instructed the station attendant to wire a report of the robbery ahead to El Paso. About the time dawn was breaking, Captain John R. Hughes and a platoon of Texas Rangers had mounted up and ridden out of Ysleta, Texas, heading toward the robbery site far to the southeast.

For three days the rangers searched the rugged and arid country south of the railroad tracks but found no sign of the bandits whatsoever. Then on the fourth day a set of tracks—three men on horseback—were found and followed only to be lost again in the exposed granite rock of the desert. For several more days the rangers circled the area trying to pick up the trail of the bandits but had no luck. Finally, Hughes called the search to a halt and, leading his platoon of rangers, returned to headquarters in Ysleta.

After leaving the site of the robbery, Ketchum and his companions rode for two days without stopping except for water and to rest their weary horses. The morning of the third day found them deep in the remote and nearly inaccessible reaches of the Big Bend country near the Mexican border. Finally they arrived at their destination, one that had been determined several days prior to the robbery—the Reagan brothers' ranch in Reagan Canyon.

Ketchum and the four Reagan brothers were long-time friends, and the ranchers agreed to hide the outlaws for a few days. Reagan obliterated the incoming tracks left by the bandits by driving a herd of cattle across them.

Ketchum remained worried that lawmen were hot on his trail, so after resting for two days he decided it was time to leave. He paid the Reagan brothers some of the silver for their help and deciding the remainder of the silver was too heavy to transport while being pursued, cached it in a shallow cave located in a gorge just north of Reagan Canyon. This done, the three men departed the area, never to return.

A year and a half later, Ketchum was captured by lawmen just after trying to rob a Colorado and Southern passenger train in New Mexico. Ketchum was shot and seriously wounded by the conductors and apprehended a short time later. After discovering the man they captured was the notorious Black Jack Ketchum, an outlaw wanted in at least four states for murder, robbery, and a host of other crimes, the lawmen lost little time in subjecting him to a speedy trial. In the end Ketchum was convicted and sentenced to hang.

While he was in prison, Ketchum cooperated with authorities relative to providing details of the many crimes he committed. During his interrogations, Ketchum admitted to the robbery of the Southern Pacific Sunset Limited near Del Rio, Texas, two years earlier. He also told of hiding out with his companions at the Reagan ranch and caching the silver in a small cave nearby. The silver, he told the authorities, was still there because he was never able to return for it.

On 26 April 1901, Thomas "Black Jack" Ketchum was hanged in Clayton, New Mexico. He left a legacy as an outlaw, one that has been captured often in books and film. Ketchum also left a small fortune in silver in a remote cave somewhere near Reagan Canyon in the Big Bend.

THE TEXAS HIGH PLAINS

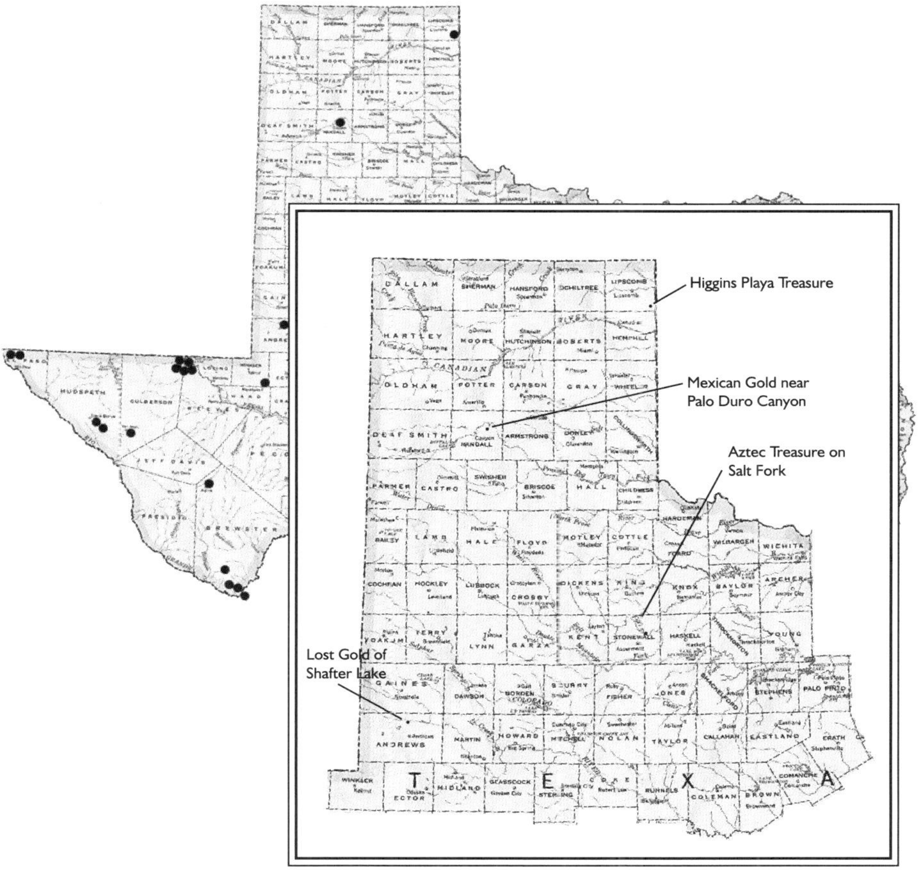

There is more to the Texas Panhandle than meets the eye. Much, much more.

From Interstate 40, the view of the Panhandle consists of seemingly endless miles of flat, arid plain. The horizon, seen dozens of miles in the distance in all directions, meets the sky. Between the observer and the horizon, little is seen sometimes save for some grasses, some cultivated fields relying heavily on irrigation, and perhaps a rare, low-growing tree or bush bent in a direction dictated by the prevailing wind.

Once the traveler departs the interstate, however, and explores the secondary and tertiary roads of the Panhandle, the environment encountered is dramatically different from that seen from the flat top of the interstate. Here and there in these high plains one finds deep, rugged stream-gouged canyons such as Palo Duro and others that fairly take one's breath away. Palo Duro Canyon is so large and so deep that it forced Spanish explorers is this region to travel several days out of their way to reach the opposite side. Once, an entire herd of buffalo was lost in the deeper parts of the canyon.

The early Spaniards called these high plains the Llano Estacado, the Staked Plains. The story goes that they pounded stakes into the ground to mark their passage so they could find their way back.

Following the Civil War, settlement in the Texas Panhandle grew very slowly at first, but then there followed a steady stream of immigrants, who eventually formed large and progressive cities such as Amarillo and Lubbock. Rich in history and culture, the Texas Panhandle is likewise rich in lore and legend, much of which is concerned with tales of lost and buried treasures.

Lost Mexican Gold Near Palo Duro Canyon

As a seventeen-year-old youth during the middle of the nineteenth century, Fergus Dooley migrated from the Indiana farm where he grew up to Central Texas, a place where he intended to find work on one of the many ranches being established in that area. He quickly found work as a cowhand and spent the next several months working with livestock.

Life as a cowboy was sometimes filled with adventure, but it was generally a poor way to make a living. Wages were low and sometimes late in coming, and economic hardship often ruled the day. As a cowhand, Dooley quickly learned that small fortunes were often and easily made by rustlers from stolen cattle and horses, so he turned to outlawry.

In a short time Dooley quit his job and put together a gang of some half-dozen experienced rustlers. During the next several months, the gang stole livestock from small farms and ranches throughout much of Central and South Texas. Though no records exist to prove it, it was often contended that Dooley and his gang killed several people.

As a result of decimating the herds throughout the region, Dooley and his gang were increasingly pursued by lawmen and vigilantes. So intense was the pressure applied by pursuers that Dooley decided it was time to go into another line of work.

One day while visiting the capital city of Austin, Dooley spotted a notice tacked to the front door of a hardware store. Pausing long enough to read it, Dooley discovered the aviso was calling for men to join the newly formed Texas Rangers. Out of money and with no prospects for any other kind of job, Dooley signed on.

In less than a month, Dooley was given command of a company of twelve recruits, mostly boys in their teens with no experience and no discipline. After working with them for only one week, Dooley was given orders to take his company to the settlement of Fredericksburg, located some sixty miles west of Austin, set up camp immediately outside of town, and provide protection to the settlers there from Indians and rustlers.

On the first day out of Austin, the young ranger company traveled about thirty miles before deciding to stop for the night. Tired and hungry after spending the day in the saddle, the rangers were eager to turn their horses loose to graze and start supper, and began looking for a suitable site.

As the rangers searched for an appropriate place to spend the night, Dooley spotted a campfire a short distance up the trail and led his men toward it. Riding into a grove of trees, the rangers encountered a group of eight Mexicans seated around the fire playing cards. Next to each of the gamblers was a stack of gold coins. As the rangers rode in, the Mexicans stood up, rifles at the ready, and faced the newcomers.

As one of the rangers spoke with the Mexicans in broken Spanish, Dooley saw two freight wagons that had been pulled behind a cluster of trees. Beyond the wagons, several horses grazed on the lush grasses that grew on the banks of a trickling stream.

Dooley dismounted and walked up to the man who was apparently the leader of the Mexican party, a tall man with a knife scar running from his hairline down across his left cheek to his neck. Dooley explained he and his men had been traveling all day and would like to camp nearby if it would not be considered an intrusion. The leader told Dooley they were welcome to share the water and grass and even invited the rangers to share their dinner.

Following the evening meal and as the rangers were laying out their bedrolls preparatory to turning in for the night, Dooley played cards with the Mexicans. During the games, he learned from his hosts that they were traders in livestock and goods, that they were returning from a recent trading expedition farther to the north, and that they were carrying back with them just over $40,000 worth of gold coins, their profit from this trip.

Dooley continued to play cards for another two hours until he lost all of his money. While he played, he could not keep his eyes off the pile of freshly minted gold coins next to the Mexican leader. Just before excusing himself to turn in for the night, Dooley decided he would kill the Mexicans and take their gold. As he walked over to the ranger campsite, he formulated a plan.

On arriving in the midst of the sleeping rangers, Dooley woke them up and told them that the Mexicans were wanted for rustling. He told his men that when the Mexicans finally fell asleep after their card game, they would sneak up on them and slay them all.

Approximately two hours before sunrise, the rangers quietly crept into the group of sleeping Mexicans clustered around the dying embers of the campfire. At a signal from Dooley, the rangers opened fire, killing all of the Mexicans within seconds.

After ascertaining all of the Mexicans were dead, Dooley instructed his men to load the bodies onto horses and take them back to ranger headquarters in Austin. He would follow them, he said, with the wagons. Several hours later when the last of the rangers rode out of the camp, Dooley undertook a search of the wagons. Several minutes later, he found a number of canvas sacks, each filled with gold coins and bound with a cord. Removing the gold-filled sacks from the wagons, he placed them into saddlebags, which he tied onto two spare horses. After waiting another hour, he mounted up, grabbed the reins to the packhorses, and rode away in the opposite direction. Fergus Dooley, with his newfound wealth, decided to flee Texas and begin life anew in Colorado as a wealthy man.

Dooley's northwestward journey from Fredericksburg toward the Panhandle of Texas was, for the most part, uneventful, and he rarely encountered travelers or settlers. By the time he entered the Panhandle and arrived at a crossing on the Double Mountain Fork of the Brazos River, he noticed he was being trailed by a dozen Comanche Indians on horseback. For several days the Indians followed Dooley without making any advances whatsoever. At night, the frightened man could not sleep for fear he would be attacked by the Comanches and his fortune taken from him. Before crawling into his bedroll, Dooley would arrange the canvas sacks in a pile next to him where he could watch over them all night long.

Several more days passed, and the Indians continued to follow Dooley. Because they made no move to harm him, he was gradually growing less concerned about the Comanches with each passing day, and he decided they were simply curious.

One afternoon Dooley arrived at the south rim of the wide, deep canyon known as Palo Duro. After riding approximately one mile in each direction and not finding a place to cross, Dooley decided to set up camp for the night and continue his search in the morning.

The following morning as Dooley finished breakfast and was loading the packhorses, he noticed the Comanches had ridden to within forty yards of his camp. He counted fifteen of them and noted that their faces were painted in a variety of patterns. Several of the Indians held bows with nocked arrows, the rest carried war lances, richly decorated with eagle feathers. And human scalps.

Frightened, Dooley slowly climbed atop his mount and, taking the lead ropes of the pack animals, began riding in an upstream direction along the edge of the canyon. The Indians followed.

All morning Dooley rode thus, with the Indians close behind him. When the sun was high overhead, Dooley stopped to fix lunch. Keeping an eye on the Indians, he brewed a pot of coffee. As he did so, the Comanches, now spread out in a line, began advancing toward Dooley. Finally they stopped approximately twenty feet away, the mounted warriors forming a semicircle around the lone man. With the steep canyon wall just behind him and the Comanches in front of him, escape was hopeless. Realizing this, Dooley poured a second cup of coffee and held it out toward a warrior he assumed was the leader of the Indians.

Only a second after Dooley made the gesture, the Indians attacked. Three of the warriors caught his horses and held them tightly while the remainder charged toward Dooley. Moments later Dooley was stretched out on the ground, a war lance through his chest and numerous arrows piercing his body. As he kicked and struggled against imminent death, another of the Indians jumped from his horse and while he was still alive, sliced Dooley's scalp from the top of his head. Another warrior stepped in and cut Dooley's throat with a stone knife. His limp body was dragged to the edge of the canyon wall and shoved into the deep gorge.

The Comanches then turned their attention to the saddlebags on the packhorses. After removing the canvas bags filled with gold coins, they sliced them open and, having no use or appreciation for the white man's money, tossed the bright, shiny objects to the ground. Several minutes more passed before the Indians finally mounted up and rode away, leaving approximately $40,000 in gold lying strewn about the floor of the Panhandle plains near the edge of the south rim of Palo Duro Canyon.

Over the years the constant Panhandle winds deposited the ever-shifting sands around and atop the scattered gold coins. In 1945 an area rancher reported finding three gold coins at a certain location close to the south rim of the canyon, but when he tried to relocate the site he could not find it.

In 1983 a photographer was snapping pictures of the canyon from the south rim when he picked up a gold coin protruding from the sand. Thinking it was probably the only one, he did not search for more. Weeks later, however, he learned the story of Fergus Dooley and the stolen gold coins and determined to return to Palo Duro Canyon to retrieve the rest of the treasure. He could never relocate the exact site where he found the single coin.

In all likelihood, Fergus Dooley's stolen coins still lie on the ground at the edge of Palo Duro Canyon where they were first scattered well over a century ago.

The Higgins Playa Treasure

For the last two centuries, a number of strange tales of buried treasures on the Texas High Plains have surfaced. Like many stories and legends of such things, the so-called Higgins Playa treasure was cached as a result of an emergency, those who buried it were forced to flee, and as so often happens, they were never able to return to recover it. The truly bizarre element of this story concerns the likelihood that some of these treasures consisted of Aztec gold, jewels, and other artifacts—treasure that was taken from Mexico and transported northward into the United States.

One of these tales concerns the caching of a large amount of gold coins along with a two-foot-tall solid gold image of an Aztec sun god near the northeastern corner of the Texas Panhandle.

The year was 1859, and a twenty-mule pack train was winding its way across the Llano Estacado. Though few details about this expedition have ever been verified, it is believed by most who have researched it that the party had been traveling for months, having left Mexico with an incredibly large amount of gold that was stolen from the country's treasury. The leaders of the pack train, as well as the muleteers, some thirty men in all, constantly scanned the horizon behind them to make certain they were not being pursued, for they well knew that the gold they carried was enough to make thirty men extremely wealthy.

At the end of one warm day of travel, one of the leaders directed the pack train toward a broad, shallow lake ringed by a few thin, stunted trees. Here they would spend a few days resting men and animals alike before continuing on to their destination, believed to be St. Louis.

As the sun was beginning to set, the herders unloaded the burros and staked them out to graze nearby. Campfires were flaring and preparations were being made for dinner when the thunder of hoofbeats shattered the relative calm of the evening.

From the south, at least two hundred Kiowa warriors were charging toward the camp, loosing arrows and lances among the Mexicans. Following the initial charge, the Indians passed completely through the campground to a point approximately one hundred yards north of the playa where they regrouped.

One of the leaders of the pack train immediately ordered several of the herders to begin excavating a hole next to the lake. By the time the Kiowas initiated their second charge, the hole was being filled with the packs of gold and the sun god image. They were refilling the hole as the Indians were riding among them, slashing and clubbing.

The second attack was devastating, leaving only five men alive. As the Indians paused before the third and what they anticipated would be the final assault on the party of Mexicans, the five men secured their mounts, saddled up, and rode away as fast as they could. Eager to rob, scalp, and mutilate the dead men in the camp, the Indians simply let them flee.

Weeks later only three of the five Mexicans who escaped from the Comanches arrived in St. Louis. What happened to the other two can only be conjectured. While in St. Louis, one of the men recorded in his journal an account of the attack by the Comanches, the caching of the huge treasure, and the escape.

During the three weeks the three Mexicans spent in St. Louis, one of them was killed when he was struck by a wagon while crossing the street. Another simply disappeared, and it was never determined what happened to him. The third booked passage on a paddlewheel to New Orleans where he intended to organize an expedition back to the small lake in the Llano Estacado and retrieve the buried fortune in gold and jewels.

There are several versions of what became of this expedition. One account relates that, as a result of fighting among the participants and the subsequent stabbing of the surviving Mexican, it never left New Orleans. Another version maintains that the expedition actually made it to the eastern edge of the Llano Estacado and was slaughtered there by either Comanches or Kiowas. Yet another version maintains that squabbling among the participants forced the expedition to break up near Wichita Falls.

Though the ultimate fate of the mission remains clouded, one important fact remains. The journal written by the Mexican was eventually found and fell into the hands of a man named John Sisney. After passing through the hands of several owners, the journal finally came into Sisney's possession while he was living in Tucumcari, New Mexico. Tucumcari is located some two hundred miles southwest of the alleged site of the treasure cache.

Following the directions found in the journal and noting pertinent landmarks indicated therein, Sisney eventually found himself in Lipscomb County, Texas, the northeasternmost county in the Panhandle. After spending two days exploring in the area, Sisney finally focused on a region near the Lipscomb-Hemphill County line and just a short distance from the Oklahoma border. Here, on a ranch owned by one George Griffith, Sisney located a small playa that satisfied all of the locational and directional criteria.

Sisney set up camp on the south side of the lake where he believed the Mexicans had camped when they were attacked. For the next several days, he spaded and raked the region in search of the cache. Near the end of the second day, his shovel turned up a handful of Mexican gold coins, each dated from the 1850s. Continued digging failed to find the cache, and Sisney concluded the coins had probably been spilled during the panicked effort to bury the treasure. Though he worked in the area for another three weeks, he failed to locate the cache and was finally forced to return home to Tucumcari.

Nagging health problems kept Sisney from returning to the playa on the Griffith ranch, and what became of the old journal is not known. Most researchers have concluded the great Mexican treasure—consisting of tens of thousands of dollars' worth of gold coins along with a large gold image of the sun god—is still lying in a shallow cache next to a playa in the southeast corner of Lipscomb County, Texas.

Amazing Aztec Treasure on the Salt Fork of the Brazos River

Most professional treasure hunters agree that one of the most amazing and puzzling, yet probably true, tales of buried treasure in Texas involves the lost Inca cache buried near the Salt Fork of the Brazos River. This treasure, believed by experts to be worth at least $100 million, consists of forty mule-loads of gold and silver ingots, emeralds described "as large as goose eggs," and hundreds of jewels consisting of precious stones set in almost pure gold and silver. What may be even more valuable, at least to historians, is the probability that the cache contains what is believed to be one of the first copies of the Holy Bible ever to arrive in the New World.

In 1531 the ruthless and bloodthirsty Spanish explorer Francisco Pizarro led a force of nearly two hundred armed soldiers into the remote regions of Peru. Accompanying the soldiers were the normal contingent of cooks, herders, and women, as well as several Catholic priests. The Spaniards, in their quest for conquest and riches, attacked and looted every village they encountered, killing men, women, and children, seizing livestock, and stealing whatever wealth in the form of gold, silver, and precious stones they could find.

After several weeks the Spaniards arrived at the large Incan village of Cotapampas in the Andean foothills. Here, at last, they found even larger quantities of gold, silver, emeralds, and other riches. With little hesitation, they attacked and seized the city, killing dozens of citizens in the process. When questioning several elders as to the source of the gold and other wealth, Pizarro was informed that the city of Cajamarca, located some six hundred miles to the northwest, was a veritable storehouse of gold, silver, and emeralds. Eager to locate this great city of wealth, Pizarro enslaved several of the Cotapampas Indians and forced them to guide the soldiers to Cajamarca.

It was around the middle of November 1532 when Pizarro and his troops arrived at the outskirts of Cajamarca, a city of some two thousand residents. Pizarro sent one of his priests along with one of the Cotapampans to serve as translator to seek an audience with the region's ruler, a man named Atahualpa.

When Atahualpa appeared, he was informed by the priest that he and his people must immediately accept the teachings of Christianity, deliver all weapons to the gate, and allow the Spaniards to take possession of the city. The priest then pressed a Bible into Atahualpa's hands.

Surprised and then angered at the demands of the strangers, Atahualpa threw the Bible to the ground and ordered the gates of the city closed to the newcomers.

The Spaniards, seasoned warriors all, wasted little time in attacking the city. Crashing through the gate, they swarmed through Cajamarca, indiscriminately clubbing, slashing, and spearing anything that moved. Though vastly outnumbered, the Spaniards were better armed and trained in warfare than the Indians, and it was only two hours later when the few hundred Cajamarcan survivors, including Atahualpa, were taken prisoner. During the ensuing week, even these poor souls were subjected to horrible torture and ultimately death.

At the end of the week, Atahualpa was among the handful of survivors. Chained at the wrists and ankles, Atahualpa, along with perhaps a dozen Indians, were brought before Pizarro and made to kneel. Pizarro then ordered the remaining Indians to travel to the outlying villages and inform them that if they did not bring in all of their gold, silver, and precious stones, that Atahualpa would be slain. For the next several weeks, the runners reached every village within one hundred miles of Cajamarca, and soon the wealth was being delivered. By the time six weeks had passed, three entire rooms were filled to the ceiling with gold and silver ore, ornaments, jewelry, and statuary, along with heavy packs of emeralds.

While at Cajamarca, Pizarro learned of a city far to the north called Quito, a city believed to have a huge storehouse of gold and silver and jewels. He quickly ordered a contingent of soldiers, along with a priest and translator, to travel to that city, seize the wealth, and return with it to Cajamarca. When the contingent of Spaniards arrived at Quito, however, the citizens mounted a resistance, and fighting broke out that lasted several days.

When Quito's treasure had not arrived within the appointed time, Pizarro ordered a dozen soldiers under the command of a trusted captain to travel to the area to look into the delay. When the soldiers arrived several days later, they joined in the fighting, and in a short time the city was taken. The captain soon discovered that, indeed, the treasure of this northern city was quite impressive. He immediately commandeered a number of llamas onto which were loaded the gold, silver, and jewels.

As the captain watched the treasure being loaded, he marveled at the vast fortune that passed before his eyes: thousands of bars of gold as well as huge emeralds set in magnificent gold and silver necklaces, rings, and earrings. The more the captain considered this incredible fortune, the more he thought about the possibilities of keeping it for himself. Finally, convinced he could be wealthy beyond his wildest dreams, he devised a plan wherein he, along with some trusted companions, would escape with the largest portion of the treasure to the north.

Along with six soldiers who were not part of his scheme, the captain sent several llama-loads of treasure back to Cajamarca, explaining that the rest of the contingent would depart the following day and meet them at the city. By the time the first pack train was well down the trail, however, the captain and his followers, along with a number of Indians who were taken as slaves, herded the second and larger pack train toward the north. Accompanying the captain was the priest, who traveled on foot carrying only a small pack of his belongings. Among his belongings was the Bible offered to Atahualpa, the first Bible, some claim, to reach the new world.

As the small party led the treasure-laden llamas northward along the ridges of the Andean Mountains, far to the south Pizarro, despite his promises to the contrary, had Atahualpa slain. The three rooms of treasure that had been accumulated was loaded onto every available pack animal they could find. This done, Pizarro led his bloodthirsty soldiers on a killing and looting rampage across much of what is now Ecuador, Colombia, and Peru, taking gold, silver, and jewels and adding to the already immense amount of riches they had accumulated. Pizarro's zest for conquest, along with his numerous successes, distracted him from the issue of the desertion of his trusted captain and the loss of a portion of the Quito treasure. As a result, Pizarro never got around to organizing pursuit and simply forgot about them altogether.

For many months the traitorous captain and his followers led the treasure-laded llamas northward through high mountain reaches, dense jungles, and arid deserts. They passed out of the mountains and traveled the narrow isthmus of Panama, up through the steamy jungles of Central America and southern Mexico, and in time crossed the Rio Grande and arrived in Texas.

During this time several of the Indians and two of the Spaniards were lost to fever, two of the slaves had been whipped to death, and the remainder of the party commonly suffered from hunger and thirst. Though their clothes had been reduced to little more than dirty rags and their boots had long since worn out, the captain kept the soldiers encouraged by telling them of his dreams of establishing a new kingdom some place in the north, a kingdom financed by the riches they carried, a kingdom in which they would all be wealthy men. When his vision of a kingdom failed to inspire them, he would warn them that Pizarro had likely sent a contingent of soldiers in pursuit. With thoughts of the horrible tortures that would be inflicted upon them by the ruthless Pizarro, the Spaniards found new strength and continued their northward march.

Throughout the hardships faced by the travelers, the captain, using a piece of tanned leather, managed to keep precise notes of their progress. Included with his notes was a detailed map. One afternoon as the party was setting up camp near a pair of low mountains, the captain made meticulous additions to his leather journal, sketching all of the nearby landmarks. Many years later it was discerned from interpreting this map that the site of the camp was near the Double Mountains in Stonewall County located between the Salt Fork of the Brazos River and the present-day town of Aspermont in the Texas Panhandle.

As the captain carefully worked on his map, one of his soldiers approached and informed him they were being watched by Indians. Looking up, the captain spotted several dozen armed warriors observing the Spaniards from a nearby ridge.

There was little sleep for the group that night, and it was a nervous party that broke camp the following morning and continued on their journey. As they traveled, they noted they were being followed closely by the Indians and that their numbers had doubled.

After passing Double Mountain on their northward trek, the Spaniards halted at the south bank of the Salt Fork. Convinced the growing contingent of Indians was on the verge of attack, the captain made a decision: They would unload the treasure from the llamas, bury it at this location, and then flee unencumbered. When they were certain the threat of attack was over, they would return, retrieve the treasure, and continue with their journey and their dreams of a kingdom.

During a time when they were not being observed by the Indians, the captain ordered the slaves to excavate twenty-one holes, each of them approximately six to eight feet deep. The treasure from the pack animals was divided and placed into the holes. As the holes were being refilled, the priest, after wrapping the Bible tightly in a piece of soft leather, placed it in one of the excavations atop the treasure. Once the holes were refilled, the pack animals were led back and forth across them, completely obliterating any sign of digging. That evening, after the Spaniards set up camp and constructed a rude fortification in case they were forced to defend themselves against an Indian attack, the captain made more entries on his piece of leather, detailing the location of the buried Inca treasure.

With the Indian threat apparently diminished for the time being, the captain decided to remain at this location for several more days. Once he was certain the Indians no longer posed a threat, he would have the treasure dug up and they would continue on their journey.

At this point the tale grows murky, and it was never determined what exactly became of the Spaniards. The record of their journey as manifested on the piece of leather ends at this point. Some researchers claim that the group continued northward and eventually succumbed to thirst and starvation. Others maintain they were eventually victims of the Indian attack they so feared. Evidence for the latter was discovered in 1887 in the form of numerous skeletons—human and llama—along with Spanish armor and weaponry near Kiowa Peak in the northeastern part of Stonewall County. The all-important leather document maintained by the Spanish captain was not found, but it appeared many years later.

Following the Civil War, the West and especially Texas opened up to settlers fleeing the war-torn South. The Panhandle region of Texas and its lush prairie grasses appealed to many, and in a short time small farming and ranching settlements sprang up across the high plains. Even during that time, a tale of long lost Incan treasure buried somewhere in the Panhandle was passed among the residents and newcomers, but no organized attempt to find and retrieve the treasure was ever undertaken.

A most interesting event occurred in 1876. An elderly Spaniard driving an expensive black lacquered carriage pulled by two blooded horses arrived in Stonewall County and immediately began purchasing parcels of land near the Double Mountain region of the Salt Fork. The Spaniard spoke to very few people, preferring to conduct business quietly and confidentially. Most residents of Stonewall County presumed the newcomer was interested in establishing a ranch, but it soon became clear that was not his intention.

Curiously, the old Spaniard erected a tent at a certain location on his property, stocked it with supplies, and moved in. From time to time he was seen walking around his holdings and consulting what appeared to be a map on a large piece of leather.

For nearly eight months, the old man lived on the property. Then, as suddenly as he appeared, he just as quickly vanished, riding off in his carriage late one night, never to return. After he left, several local residents rode out to his holdings and looked around. What they discovered was indeed strange. Here and there at one location they found buckets, shovels, and other tools lying about as if suddenly abandoned. Even more curious, they discovered a total of ten holes, each one approximately six to eight feet deep. It was clear from the markings found inside the holes that they once contained packs and crates.

Those familiar with the tale of buried Inca treasure in the area were convinced the old man must have found it, or at least some of it, and carried it away. No one ever discovered his true identity or where he went. What was quickly discerned, however, was that there remained eleven holes unexcavated. Presumably, they were near the ten holes dug by the Spaniard, but a few random excavations yielded nothing.

During the ensuing years, floods from the Salt Fork have filled the ten excavations with sediment. And while there are some old-timers in the county who claim to know their exact location, there exists some confusion about the actual site of the burial of the incredible Incan treasure nearly five hundred years earlier.

Strange indeed, but events were about to get even stranger.

In the spring of 1902, a second Spaniard arrived in the area, this one a young man in his late twenties, apparently well educated and, oddly, carrying a very old map drawn on a large piece of tanned leather! Many speculated as to the identity of the stranger, but like the Spaniard who preceded him two and a half decades earlier, this one likewise remained quiet and elusive.

After spending several days riding around the property purchased by the previous Spaniard, the younger one rode into Aspermont, purchased a number of shovels, picks, and buckets, and hired four men to do some digging. One of the men, who was interviewed several years later, stated that the young Spaniard, constantly consulting the leather map, led the diggers around the area for four or five days checking landmarks and asking them questions about the region. Finally they arrived at a location decided upon by the young man and proceeded to dig.

Within the first hour of digging, the workers unearthed four skeletons. They were then directed to excavate other holes at marked locations. Several of these holes appeared to have been previously excavated and then filled in with river sediment. Though they dug as much as ten feet into the ground, nothing was ever found.

On the day they were due to be paid, the Spaniard explained he did not have enough money but would share a percentage of the treasure he claimed they were certain to find. Following several more days of digging in the hot Panhandle sun, the laborers grew angry and finally demanded their wages. The Spaniard assured them he would go into town and obtain some money from the bank. After carefully folding the leather map, he mounted his horse and rode away. It was the last anyone ever saw of him.

Most researchers who have studied this tale believe the young Spaniard was related to the previous one, perhaps a son or a grandson. They are also convinced that the reason he did not find any of the buried Incan treasure is because the workers dug into the same holes previously excavated by the elderly Spaniard years earlier.

As far as anyone knows, the remaining eleven caches have never been found and remain filled with a great portion of the incredible Incan treasure.

The Lost Gold of Shafter Lake

Shafter Lake, located in Andrews County just a few miles north and west of the city of Andrews, is not so much a lake as it is a sometimes water-filled depression located in this semiarid environment on the Texas High Plains. Such depressions, many of which dot the Texas Panhandle, are called playas and generally gather saline-laden water from the runoff generated by infrequent thunderstorms. Intense heat and low humidity facilitate evaporation, which in turn leads to the drying up of most or all of the water in these basins and the formation of a crusty accumulation of salt. Shafter Lake is often dry much of the year.

In addition to being one of the few part-time sources of moisture in this part of the Texas High Plains, Shafter Lake is also associated with one of the richest and most mysterious lost treasures in the United States.

William Rufus Shafter enlisted in the Union army at the outbreak of the Civil War. By the time he was twenty-six years of age, Shafter had seen action at Ball's Bluff, Fair Oaks, Nashville, Yorktown, and West Point. Twice wounded and often decorated, Shafter won the recognition and praise of General George Henry Thomas, the so-called Rock of Chickamauga. As a reward for Shafter's contribution to the war effort, Thomas placed him in command of a large contingent of black troops assigned to frontier Texas. In a very short time, Shafter earned a reputation as a fearless and indomitable commander who would ride into battle at a moment's notice.

During his West Texas assignment, he picked up the nickname "Pecos Bill."

In 1957 a researcher accidentally found some rather cryptic and undocumented military reports that strongly suggest Shafter and a contingent of black cavalry troops were escorting two wagonloads of gold from the Mexican border to some unidentified location in the United States. Where the gold came from could not be learned from the documents, and its source remains a mystery to this day. Furthermore, the total amount of gold being transported is not known, nor is it known what form it was in—ingots, coins, or nuggets. A search for substantiating documents has proved fruitless, but most researchers conservatively estimate that "two wagonloads" of gold today would likely be worth well over two million dollars.

The lack of information regarding this gold shipment has led a number of researchers to consider the notion that Pecos Bill Shafter may have been illegally transporting the gold and that he intended it for his own use. Passionate Shafter scholars who resent the impugning of the soldier's integrity have suggested it is more probable that he was on a secret mission for the United States government and that the gold was likely on its way to the U.S. Treasury.

Whatever the truth, and it may never be known, the gold never arrived at its intended destination.

As the tale unfolds, Shafter, leading his party of mounted troopers and the two wagonloads of gold, had just passed to the east of Monahans, Texas, and were traveling northeasterly when one of the scouts spotted a band of forty to fifty Comanche Indians trailing a short distance behind. Shafter was well aware of the danger posed by these plains raiders, for he had been sent into this part of Texas to protect travelers and settlers from the depredations of the hostile tribes. Concerned, Shafter assigned a half-dozen sharpshooters to ride at the rear of the small caravan.

Throughout the day the Comanches remained approximately one mile behind the party of soldiers, never advancing or making any threatening gestures. Just before sundown the soldiers arrived at a small spring located not far from the present-day town of Notrees, and Shafter instructed the men to set up camp. Few of the black troopers slept that night as they kept watch for the Comanches they knew were just a short distance away.

Early the following morning after a cold breakfast, the party set out on the trail once again. As the rising sun illuminated the bright, arid plains, the soldiers glanced backward to see the Comanches riding behind them.

Travel was growing increasingly difficult through this part of Texas, and great effort was often expended negotiating the heavy wagons through sand-choked trails and across deep washes and gullies. Two days later the slow-moving contingent finally made camp not far from the southwestern shore of a large playa.

Several men hauled water barrels out to the shallow lake to fill them but found the water too salty to drink. They contented themselves with a bath, the first they'd had in several days, and noticed the water was approximately three feet deep.

That evening as the troopers ate their supper, Shafter and one of his lieutenants studied the large playa and decided, rather than lose time by traveling around it, to simply drive the wagons and men across it in the morning.

The following morning the men dined once again on a cold breakfast, hitched the mule teams to the two gold-filled wagons, and readied their own mounts and equipment. A short time later Shafter signaled the troopers forward into the playa.

The initial phase of the crossing was uneventful. The saline waters lapped at the sides of the heavy wagons, but progress was not slowed. As the party approached the center of the lake, however, trouble started. The lead pair of mules became bogged in the softer mud of the bottom and were unable to continue. As the soldiers worked to free the mules, the trailing pair began to sink, soon to be followed by the wagon itself, slowly but surely disappearing into the saturated sands and silts at the bottom of the lake. As the top of the first wagon sank below the surface of the water, the second one, along with the mules, became bogged down and was also sinking. Extra horses were harnessed to the wagons and mules but to no avail—they too became mired in the mud and drowned.

As Shafter watched, stunned at these developments, the Comanches rode into view from the south and, forming a single line along the shore of the playa, watched the activities of the troopers. When it became clear that the soldiers had been placed in a somewhat helpless and near defenseless position in the middle of the lake, the leader of the Comanches screamed out a loud war cry, and suddenly the mounted and armed Indians swept into the playa and rode toward the confused troopers. Calling for his men to abandon the wagons and mount up, Shafter led the disorganized party of cavalry to the far shore. Without stopping, they continued their flight from the attaching Comanches across the dry plains to the north.

The Indians chased the troopers for approximately four miles before reining up. Returning to the wagons mired in the lake, they inspected them but found little useful save for some ropes, harnesses, and canvas. Satisfied they had obtained any important items, the Comanches returned to the south from whence they came.

During this time, this part of Texas was seldom visited by travelers or settlers. As a result the two wagons partially sunk into the floor of the playa went undiscovered. Over time, the wagon planks rotted away, spilling the shipment of gold onto the soft bottom of the lake. The heavy weight of the gold quickly sunk out of sight into the saturated sands and silts of the playa.

According to existing military records, Shafter and his command never returned to this region of Texas. Therefore, it is reasonable to believe that the huge shipment of gold was never retrieved by the only persons who were aware of its location. Shafter was given other assignments, one of which was to lead a company of cavalry during the Santiago Campaign on the island of Cuba in 1898. William Rufus Shafter died in 1906.

Sometime during the year 1901, the rotted wagons that carried the fortune in gold into the middle of the Panhandle playa were found by one William Russell.

Russell, along with his wife and three children, were traveling from Denton, Texas, to Pecos when their wagon broke down northeast of Andrews near a dry lake bed. While Russell and his wife made repairs to the wagon, his sons explored the crusty surface of the playa. During dinner that evening, Russell noticed one of his sons playing with some items he had never seen before—pieces of a military harness and some wagon fittings. When he asked him where he obtained them, the boy pointed out to the middle of the playa and said he found them lying on the surface. The following morning before commencing work on his wagon, Russell walked out onto the playa and discovered the rotted and rusted remains of at least two wagons. He also found bones of what he believed were horses and mules. As Russell searched the area, he was, of course, completely unaware that some short distance beneath the surface reposed an incredible fortune in gold.

Russell finished the repairs to the wagon and continued on to Pecos where he established a relatively successful commercial orchard and garden. Some ten years after arriving, he told a friend about finding the wagon parts out on the playa not far from Andrews. The friend then related a story he once heard about Shafter's lost treasure and how he was forced to abandon it in the lake bed just outside of Andrews.

Excited at the possibility of locating the rumored treasure, Russell, along with two of his sons and the friend, traveled to the playa—now called Shafter Lake—to try to recover the gold. On arriving, however, they discovered the lake had been filled to a depth of three to four feet as a result of recent rains, and they had no success in locating the site of the wagons.

In 1931 a ranch hand told a story in Andrews that, while out searching for some stray cattle about two weeks earlier, he was crossing the "old lake bed" when he found several pieces of an old wagon. Two ranchers who were familiar with the tale of the Shafter Lake treasure went out to the playa the following day, but found most of the lake bed covered by sands blown in by the winds.

The bed of Shafter Lake has been studied several times by professional treasure hunters. It has long since been determined that when the playa fills with water, the porous soil expands, providing for the easy sinking of heavy particles or objects. While virtually all who have searched for the Shafter Lake treasure are certain that the gold has sunk into the bottom soils, there is little agreement as to how deep it may have proceeded, with estimates ranging from two feet to nearly ten feet.

A number of searchers have employed metal detectors while exploring the lake bed during periods of drought. While a few wagon and harness parts have been found, the location of the gold remains unknown.

(During the writing of this book, it was learned that a recovery company was in the process of securing the rights to excavate a portion of Shafter Lake in search of the gold.)

The Rest of the West

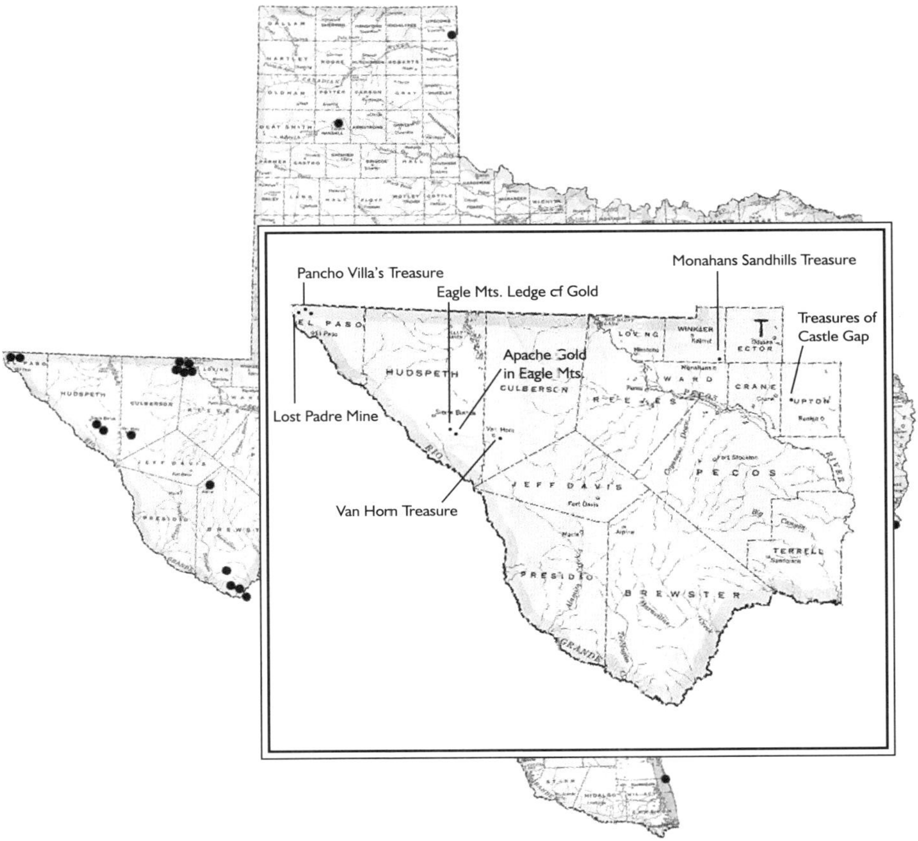

While the Guadalupe Mountains and the Big Bend region of Texas seem to dominate when it comes to being settings for tales of lost mines and buried treasures, there are other locations in the vast, broad, and expansive West Texas region that have yielded no less exciting stories.

When one pauses to consider the landscape and environment that is West Texas, one becomes quickly impressed with the great variety found in the region. In addition to the tall peaks and steep slopes of the Guadalupe and Chisos Mountains, one can also find the Davis Mountains, the Franklin Mountains, and a number of lesser ranges.

In addition to these rocky environs and higher altitudes, one can find hundreds of square miles of Chihuahuan Desert with its population of unique and specially adapted plant and wildlife populations. Ranging northward into the Panhandle, one encounters what the Spanish explorers of centuries earlier called the *Llano Estacado*, the staked plains, a region of seemingly endless prairie that stretches in all directions to the horizons. Once into this appealing high plains region writer Dan Flores calls the Horizontal Yellow, however, one finds surprisingly deep canyons and gorges, some large enough to hide an entire city or vast herd of buffalo.

Aside from the aforementioned mountains and related features, much of the rest of West Texas yields little or no outcrops of ore in the form of gold or silver, thus mining activity was for the most part absent. The transport of ore and other manifestations of wealth across these vast stretches of the plain and prairie and desert, however, were common. Many of these treasures had to be abandoned to facilitate travel, some were stolen and cached, and some were merely lost.

All were added to the great store of fascinating tales and legends of lost treasure.

The Lost Padre Mine of the Franklin Mountains

In far West Texas and lying immediately north of the city of El Paso are the Franklin Mountains, a north-south oriented mountain range containing an impressive diversity of rock, ranging from volcanic to sedimentary. Somewhere in this range, someplace in some long-forgotten canyon location lies one of the oldest and richest gold mines in the history of North America, one that is associated with the Spanish priests who once held great influence over the inhabitants of this region, and one that continues to mystify treasure hunters to this day.

In 1659, in the town of El Paso del Norte, now Juarez, Mexico, Catholic Spanish padres oversaw the establishment of a mission, the *Nuestra Señora de Guadalupe*. The mission stands today near the center of Ciudad Juarez, a city with well over one million citizens. During the seventeenth century, however, this area was sparsely populated and saw the passing of trappers, prospectors, soldiers, and clergy on their way to outposts located far to the north. The mission, as well as the adjacent buildings, was constructed with Indian labor. The Indians generally came from the peaceful populations that were occasionally found in the area living along the river bottoms, and a number of historians maintained they were enslaved to perform work for the priests. Furthermore, scholars claim, the Indians were routinely whipped and starved, and the mortality rate among them was very high.

Every morning, both during and after the construction of the mission, a small group of priests, accompanied by a number of Indians, crossed the Rio Grande, made their way into a canyon in the Franklin Mountains, and proceeded to mine rich gold from a location they had discovered earlier. When a sufficient amount of gold was accumulated from the mine, the Indians would carry the ore back down the canyon and to a crude arrastre located on the banks of the river where it was separated from the rock matrix, melted, and then poured into molds to form ingots. Each time a few hundred of these ingots were accumulated, which was approximately once every month, they were loaded onto burros and transported to the church headquarters in Mexico City.

Legend has it that every morning one of the priests entered the mission tower to ring the bell summoning all the other priests to worship. From this vantage point, it is said, one could look north across the river and, in the light of the rising sun, see the opening to the padres' gold mine.

The activities associated with the operation of the mission and the mining of the gold continued apace for several years, and during that time virtually all of the Indians in the region were converted to Catholicism, sometimes voluntarily, sometimes by force. The church prospered, not only in terms of converts, but in material wealth as well.

One afternoon in the year 1680, a messenger arrived at the mission, having ridden a long distance from the north, and the news he carried was not good. The Pueblo Indians who resided many miles to the north had grown tired of enslavement by the priests and rebelled against the strict authority of the Catholic Church. Unable to endure the harsh rule of the Spaniards any longer, the Indians revolted, killing many priests and soldiers in the process and driving away the survivors. The few who escaped, including the messenger, were warning others as they made their way south and into Mexico. The Pueblo Indians, under the leadership of Chief Cheetwah, were sweeping southward and killing all Spaniards with whom they came in contact.

Concerned they were in the path of the rampaging Pueblos, the priests of the *Nuestra Señora de Guadalupe* gathered all of the church valuables, including chalices, vessels, candlesticks, platters, golden ornaments of various kinds, along with hundreds of gold ingots that were waiting to be shipped to Mexico City. After loading this great wealth onto mules and burros, it was transported to the mine in the Franklin Mountains, 250 loads in all according to the legend. There, it was carried into the shaft and stacked against the walls. From the river, Indians and priests alike carried bucketloads of red river silt up to the mine and used it to fill, it is said, the entire shaft. According to the legend, three of the priests were killed and buried in the shaft, presumably so their spirits could guard against the possibility of anyone but Spaniards recovering the gold. Following this, great care was taken to stack rocks in front of the mine in order to disguise the entrance to make it look like the rest of the canyon.

For two years the Spaniards alternately fought with and fled from the warring Pueblo Indians. Little time was available for running a mission, converting pagan souls to Christianity, or operating a gold mine. Many of the priests who worked or administered the mine fled to Mexico City, leaving the mission solely in the hands of subordinates.

Eventually the Pueblos were subdued, and the Spaniards gradually drifted back into the region to reclaim what would become New Mexico and a portion of Texas. On arriving at the mission *Nuestra Señora de Guadalupe*, the church officials made plans to reopen the mine, return the church valuables to the mission, and commence mining the gold once again. However, when they entered the canyon in the Franklin Mountains, they could not relocate the mine! Most of the original priests who worked the mine were dead, and the remainder had been transferred back to Mexico City or to Spain.

Though the new priests searched for the mine for several months, they were unable to find it. Finally they gave up and returned to their duties at the mission.

During the early 1880s, a researcher working in the Catholic Church archives in Mexico City came across a reference to a very rich gold mine once operated by the Spanish priests and located in the Franklin Mountains just north of El Paso. Continued research yielded information on how many burro loads of gold had been transported from the mission *Nuestra Señora de Guadalupe* to church headquarters in Mexico City. The researcher also encountered other documents, including several that detailed the hiding of 250 burro loads of church valuables and hundreds of gold ingots in the Franklin Mountain mine around the time of the Pueblo uprising. Fascinated at the possibility of so much wealth, the researcher organized a group of investors to search for the mine, recover the golden church artifacts, and resume mining the rich gold purported to be found within.

The investors hired several men to search for the gold, including a mining engineer, a geologist, and a hydrologist. Despite spending several months exploring the various canyons in the Franklin Mountains, they found nothing.

Having heard the legend that one could stand in the bell tower of the mission in Juarez and see the opening of the gold mine, a man named Robinson did exactly that. From his position in the tower, Robinson located two or three likely canyons in which the mine might be located. After spending time searching the area, Robinson found evidence of a very old trail that led from one canyon back down the slope to the river and eventually back to the mission. Recognizing this important clue for what it was, he invested more and more time searching throughout this particular canyon.

Along one side of the canyon, Robinson detected a pile of rock and debris that did not appear to be a natural deposition—it looked as though much of the rock consisted of mine tailings mixed with some larger stones. Curious, he laboriously removed much of the rock and was surprised to discover a rock- and dirt-filled opening to a mineshaft. The dirt was a type of red silt that had clearly come from the flood plain adjacent to the nearby Rio Grande.

Recognizing the incredible amount of work facing him, Robinson returned to El Paso, sought for and secured a small amount of financial support, and hired half a dozen laborers to dig the dirt from the shaft. Several yards of the silt had been removed when Robinson's supporters withdrew their funding. Angered at this development, Robinson had his workers refill the shaft and cover it over once again with rocks and debris so no one else would be able to find it.

What eventually became of Robinson is not known.

In 1901 an El Paso newspaper reported that a man named L.C. Criss had found what has come to be called the Lost Padre Mine. According to the article, Criss had been searching for the mine for nearly a decade and a half based on some information he encountered in an old manuscript he found in the mission library. Criss claimed that from the mine, one could climb to the adjacent ridge and look down onto the city of El Paso.

Like Robinson, Criss found the jumble of rocks, removed them, and located the shaft filled with river silt. Financing the project himself, Criss removed enough silt to expose 125 feet of the shaft. On reaching approximately 100 feet into the shaft, Criss and his workers began finding ancient Spanish tools and artifacts, including spurs, an anvil, spikes, and a rotted boot. For a time these items were placed on display at an El Paso business and were seen by thousands.

When Criss and his laborers had arrived 125 feet into the mine, they discovered that the shaft forked, with each fork going in opposite directions like the cross bar of a "T." Interestingly, the opening to each of these shafts was walled up and tightly cemented with adobe bricks. Selecting one of the shafts, Criss tore out the wall and removed at least another twenty feet of red river silt before he noticed that the timbers supporting the ceiling were rotted and coming apart. Concerned about the potential of a cave-in, Criss took two of the workers and drove a wagon to El Paso to purchase new timbers. Before he left the mine, he gave orders that no one was to enter the new shaft.

Once Criss was gone, however, one of the diggers, a man named Cruz, was convinced he was close to the long lost treasure of the padres and resumed digging. Cruz was in the shaft no more than twenty minutes when it collapsed, killing him instantly and effectively sealing off the passageway.

Criss was dramatically affected by the death of his employee, and he immediately cancelled further work on the shaft for several weeks. During that time, unfortunately, he exhausted his funds, was unable to procure more, and was thus forced to abandon the project. Dejected, Criss covered up the mine entrance, it was said, by generating a small landslide, and left El Paso. Like Robinson, Criss was never heard from again.

Though occasional unsuccessful expeditions were conducted into the canyons of the Franklin Mountains in search of the Lost Padre Mine, very little of significance occurred until 1968 when a man named Martin, along with his wife and an assistant, arrived in El Paso from California with documents he claimed identified the precise location of the Lost Padre Mine. The documents, said Martin, also contained a complete inventory of the church valuables that were hidden in the shaft during the Pueblo Indian revolt of 1680.

Martin made arrangements to lease a large backhoe and some other digging equipment and proceeded to the location on the flank of the Franklin Mountains indicated in his documents. With hundreds of El Paso citizens looking on, as well as television cameras and newspaper reporters recording the event, Martin directed the excavation of a portion of a Franklin Mountain slope.

The machinery gouged a long, fifteen-foot-deep trench, which came very close to a well-traveled paved road. Hundreds of tons of rock were removed and pushed downslope. At one point Martin halted the excavation and climbed into the trench to inspect it. Using a pickaxe, he worked on one section of the wall until he finally removed enough debris to reveal the outline of a man-made mine shaft, one that was filled with red dirt! Before the day was over, approximately ten yards of the shaft had been excavated.

Armed guards patrolled the trench that evening, keeping everyone away from the promising-looking shaft. In the morning just as work was about to commence, however, an official from the city of El Paso arrived, provided a quick inspection of the excavation, judged it to be hazardous to nearby residents, and ordered it halted immediately. Martin pleaded to be allowed to continue the excavation but it was to no avail. He was ordered to refill the trench and remove himself and the equipment from the area. Thwarted, Martin simply abandoned the project and returned to California, never to return.

At this writing, no one has attempted to reopen the trench and continue the excavation of the dirt-filled mineshaft.

According to professional treasure hunter and author W.C. Jameson, more than three hundred attempts have been made to locate and reopen the Lost Padre Mine. None has been successful. Recent plans to make another attempt were discarded when it was learned that homes have been constructed adjacent to and above the area where the mine is believed to be located. Any significant excavation in the area, it was determined, could potentially jeopardize the structural integrity of the bedrock upon which these homes were perched.

The fact is, gold has indeed been found in the Franklin Mountains, and on several occasions. There are people living in El Paso today who, during the 1930s, actually panned gold from the slopes of the mountain range during and after the rare thunderstorms and subsequent runoff. Several residents of the Franklin Mountain foothills have found small gold nuggets in their backyards while gardening or working on their lawns.

In addition to gold, silver has also been discovered in the range. During the early 1870s, a mining company that was prospecting throughout these mountains found two very old mine shafts, each about 100 feet deep, that had been filled with dirt similar to that associated with the Lost Padre Mine. After the dirt was removed, the miners were surprised and delighted to discover a three-foot thick vein of silver in both shafts. The shafts, noted one of the miners, were only about two miles from the old Spanish mission in Juarez. Given the proximity to the old mission and the fact that the shafts were quite old, many considered that the Spanish padres might have mined silver as well as gold in the Franklin Mountains. No record of silver mining, however, has ever been found.

In recent years other old shafts have been discovered in the Franklin Mountains, passageways that were very old and clearly the result of human labor. In some of these shafts, tools of Spanish origin have been recovered, but to date no gold or silver has ever been found, nor have any valuable church artifacts or gold ingots.

The legend of the Lost Padre Mine, now stronger than ever, continues to lure treasure hunters and adventurers to the Franklin Mountains, all arriving with the hope that they will be the ones to learn the secret of this incredible lost and buried fortune.

Pancho Villa's Treasure

The Franklin Mountains are the setting for more than one treasure. In addition to the fabled Lost Padre Mine as well as several others, many researchers are convinced this range also contains a vast fortune hidden during the early 1900s by the famous Mexican revolutionary leader Pancho Villa.

Villa, one-time bandit, horse and cattle thief, and well-known desperado, always fought for the poor and downtrodden *campesinos*. Following the outbreak of the Mexican Revolution, he was given a commanding position in the country's military and made the leader of thousands of troops.

During his days as a young bandit, Villa developed distrust for paper currency and a great fondness for gold. Whenever he learned about a possible shipment of gold emanating from some mining operation in the Sierra Madres, he generally made plans to rob it. Once, on learning that $120,000 worth of gold bullion was being shipped on a Mexican Northwestern train, Villa and his band blocked the tracks near Santa Isabel, boarded the train, and took it. When he kidnapped merchant Frank Knotts in 1918, Villa refused a $20,000 ransom offer in currency and demanded the payment be made in gold. It is claimed that when Villa staged a robbery of the Santa Rosalia Mines, he needed fourteen mules to carry all of the gold.

Over time, Villa accumulated impressive quantities of gold in the form of ingots, coins, and nuggets. He would have it strapped to the backs of mules and carried along during his raids. Eventually the bandit realized this growing pack train of gold was becoming difficult to manage and was slowing down the movement of his raiders. As a result, he began caching it in a variety of locations in the Sierra Madres. In addition to the gold, Villa's fortune consisted of purloined jewels and even golden church artifacts.

Realizing that the day might eventually come when he would be forced to flee from Mexico, Villa decided to cache a significant portion of his gold in the United States, and he chose a location in the Franklin Mountains just north of downtown El Paso, Texas. From a command post in Durango, Mexico, some 500 miles south of the Rio Grande, Villa ordered one of his trusted lieutenants to select six men to load and escort a quantity of gold to a preselected location in the Franklin Mountains. Two leather packs, each containing an immense quantity of gold, were strapped to the backs of several mules. Accompanied by the heavily armed escort, the gold-laden mules were led northward to the United States and the Franklin Mountains.

With such a heavy cargo, the journey was agonizingly slow, and it was several weeks later when it finally arrived at a river crossing just downstream from the small town of Zaragoza, located twenty miles east of Juarez and El Paso.

After crossing the Rio Grande, the pack train and escort skirted the mission town of Ysleta to the west and proceeded toward the Franklin Mountains that loomed in the distance. Using the map and directions supplied by Villa, the lieutenant found the precise canyon indicated by his leader and led the pack train up the steep trail.

Near the head of the canyon, the lieutenant ordered a halt and the soldiers began unloading the heavy packs from the mules. After consulting the map and pacing off a prescribed distance, the lieutenant marked a spot and ordered the soldiers to dig a deep hole in the bare, rocky soil.

By late afternoon a deep hole had been excavated and the packs of gold placed within. When the last bag was tossed into the pit, the soldiers grabbed their shovels and began refilling the excavation. With the soldiers' backs to him, the lieutenant walked over to his mount, pulled a rifle from the scabbard, and with no emotion whatsoever, shot and killed each of the six men. After kicking their bodies into the hole and on top of the treasure, the lieutenant shoveled the remainder of the dirt over them. On a nearby rock, as per Villa's instructions, the lieutenant scratched the word "*ORO*," Spanish for gold. This done, he remounted, rode out of the canyon, crossed the river, and began the long journey back to Durango.

Weeks later the lieutenant rode into Villa's camp and reported that his assignment had been carried out. Villa thanked him and saw to it that he was fed and his horse properly cared for. That night, however, the lieutenant was awakened by three men who entered his tent. He was tied and gagged and, in the dark of night, led out into the foothills beyond camp where his throat was slashed. Now, the only man alive who knew where the fortune in gold was buried in the Franklin Mountains was Pancho Villa himself.

Villa, however, never had an opportunity to travel to the Franklin Mountains to retrieve his gold. Before he could do so, he and his entourage were ambushed in the town of Parral, and he was assassinated.

In 1932 an American named C.F. Degner was doing business in Mexico when he accidentally found a large package of important papers that had apparently once belonged to Pancho Villa. In the package were a number of letters of a political and military nature, a description of a parcel of land that was given to the former general, and papers on a variety of other matters. Most astounding of all were several papers containing detailed directions and maps pertaining to a number of locations where Villa apparently cached different treasures. One of the locations indicated in the material specifically identified a large cache in the Franklin Mountains north of El Paso.

After securing funding and organizing an expeditionary team, Degner, using the information in the documents, led three successful searches for buried Villa treasure in Mexico's Sierra Madres. Before he could undertake the search for the Franklin Mountains cache, however, Degner passed away. What became of the papers containing the information about Villa's treasures is not known.

During the 1940s a man who had grown up in El Paso and had always heard that one could occasionally find gold in the Franklin Mountains, walked up a canyon one morning searching for promising outcrops in the rock. Near the head of the canyon, he found a place where recent runoff from the infrequent rains that fall on this region had eroded out a shallow trench across a short section of the canyon. In the hole he spotted several bones. On closer examination the man determined the bones were human. With the bones he also found remnants of clothing, some buttons, and even a piece of a rotted leather boot.

The man determined this spot must be a grave. After removing the boot and some buttons, he covered it up and rolled several boulders on top of it to inhibit further erosion from runoff. Then looking around the canyon a bit more, he gathered up the boot and buttons and returned home.

Several years later he asked a friend to look at the boot and buttons. To his astonishment, he learned that they were part of a uniform worn by Mexican revolutionary soldiers.

Ten years after discovering the origin of the boot and buttons, the man heard for the first time the tale of the caching of Pancho Villa's fortune in gold in the Franklin Mountains along with the subsequent execution and burial of the soldiers involved. Realizing that what he thought was a grave was actually the location of an incredible fortune in buried gold, he decided to return to the canyon and dig it up.

With the passage of so much time, however, he had difficulty remembering exactly which canyon was the one in which he hiked that long ago day. After selecting one in which he believed he would find the gold, he spent weeks searching it for the spot over which he rolled the boulders but found nothing that resembled it. He searched the adjacent canyons, all with the same result. In all, he spent three years looking for the site of Villa's remarkable gold cache but never found it.

In the early 1970s several members of a geology class at the University of Texas at El Paso, given the assignment of identifying certain minerals and fossils in the range, were exploring a canyon in the Franklin Mountains. In addition to notes and lists they were compiling, the students also took photographs, all of which were to be turned in to their professor.

Several weeks later geology professor William S. Strain was reading the materials submitted by the students, making notes on them, and applying grades. While examining one particular set of notes, he spotted an attached photograph that depicted a quartz seam cutting through an igneous intrusive exposure. In the corner of the photograph was a rock, as he described later, "about the size of a bean bag chair," on the face of which was inscribed the word "ORO"!

Being familiar with the tale of Pancho Villa's buried treasure in the Franklin Mountains, Dr. Strain grew quite excited about the prospect of finding it and decided to contact the student immediately. Since it was spring break and the campus was virtually empty, it was several days before Strain could obtain information on the student's address and telephone number. When he called, however, there was no answer.

When classes resumed after spring break, Strain noted that the student, a female, was no longer enrolled in class. When he inquired, he discovered she had been forced to drop out of school and return to her home in South America to care for her invalid mother. Attempts to contact her in South America were unsuccessful.

Dr. Strain realized there was only one thing to do—he would have to go to the canyon and search for the rock himself. At the first opportunity, he made arrangements to do so and spent an entire Saturday afternoon exploring the region looking for the inscribed rock and the quartz seam.

On his third trip into the canyon, Strain found them. It was very late in the afternoon and only a few minutes before sundown. Knowing he needed to get out of the canyon before dark, Strain hurriedly paced off twenty feet from the ORO-inscribed rock and examined the area of the circumference. One location in particular, he said later, looked very promising. And as he hiked out of the canyon, he made mental preparations to return the next Saturday and attempt an excavation to get at the treasure in gold he was convinced lay only a few feet below the surface.

It was not to be. Long in ill health, Dr. Strain died only a few days after his discovery.

Most researchers who have studied the tale of Pancho Villa's buried gold in the Franklin Mountains remain convinced it is still there. Given what is known or at least suspected, most agree that the hoard would be worth several million dollars if recovered today.

The Van Horn Treasure

Van Horn, Texas, lies some 130 miles east of the Franklin Mountains along U.S. Interstate 10. Van Horn is an unpretentious town lying in a shallow basin between the Van Horn Mountains, the Sierra Diablo Mountains, and the Apache Mountains. Located in the northern part of the Chihuahuan Desert, the region is relatively devoid of vegetation save for a few arid land varieties and the occasional cottonwood or willow that grows along the banks of normally dry stream courses. When the West Texas wind blows in Van Horn, which it does often, it is not uncommon for the moving sand to reduce visibility to only a few dozen feet.

Less than ten miles east of Van Horn and located on public school lands is a strange hole, one that is known of by only a few people. Even those who know about it often have difficulty finding it in this expanse of desert. The tiny opening, only slightly more than a foot-and-a-half in diameter, serves as an entrance to a rather narrow, vertical passageway that extends downward for about eight feet. At that point it widens somewhat and continues horizontally in an eastward direction for another two-dozen feet. At the end of this portion of the passageway, there is a vertical drop that some claim is bottomless and some claim was once measured at approximately 400 feet. From this deep, elongated chamber, several branching passageways are discernible at various levels, each one extending in different directions.

One autumn afternoon in 1974, seventeen-year-old Mauricio Suarez was playing in the desert east of Van Horn when he discovered the hole. Spotting a three-foot-long rattlesnake coiled in a small cluster of desert shrub,

Suarez decided to evict the reptile from its hiding place in order to get a better look at it. To accomplish this, he set the bushes afire. The heat from the fire caused the snake to slither away and seek a better hiding place. And as the bushes burned away, Suarez spied a narrow opening in the ground among their ashy midst.

Squeezing into the narrow opening, Suarez cautiously lowered himself to the bottom of the first level and explored the lateral passage. On encountering the deep hole at the end of it and realizing he could proceed no further, Suarez decided to return to town to obtain a long rope and to enlist the help of some of his friends.

The following weekend Suarez and four companions returned to the location. After making their way through the cave to the great drop, the boys tied off the rope, and Mauricio, carrying a flashlight, climbed down to investigate the lateral passageways. He managed to climb into one, crawled on his hands and knees for several feet along the narrow tunnel, and made a most unexpected discovery. There, against one wall of the passageway, was a pile of luggage—leather satchels and suitcases. Among the dust-covered cases were also a number of cloth bags, all of them stenciled with the year 1903. Opening some of the bags and suitcases, Mauricio's heartbeat quickened as he inspected the contents. Lying there before him were thousands of bound twenty- and fifty-dollar bills, more money than he ever imagined existed. Excited, Mauricio returned to his friends, informed them of his discovery, and together the five boys made plans to return the following week to retrieve it.

The next Saturday afternoon, the young men crawled back into the cave, secured the rope, and for the second time Mauricio made his way to the lateral passageway in the deep shaft that contained the money. Just as he attempted to climb into the passageway, his foot slipped, and with only a tentative grip on the rope, he slid downward before finally landing on a narrow rock projection some fifty feet below.

Badly injured, Suarez was unable to climb back up the rope. High on the ledge above, his companions called for him to tie the rope around his torso and they would try to pull him up. Though they made a valiant effort and struggled for nearly two hours, they were unable to raise their companion because of the cramped space in which they were forced to work. They had only one choice—they had to drive to Van Horn for help.

On being alerted to the mishap in the remote cave east of town, Culberson County Sheriff Richard Upchurch, along with a deputy and two rescue workers, hurried to the site.

After trying for another two hours to raise Mauricio out of the chamber, they were no closer to rescuing him from the pit than the boys. In desperation, Upchurch called in a second rescue unit, this one comprised of professional cavers from Carlsbad, New Mexico. Around midnight a total of fourteen people arrived in the area and formulated a plan to bring the youth out of the cavern. Working all through the night, they finally hauled Mauricio to the surface at dawn of the following day. The boy was exhausted, frightened, scratched, bruised, and cold, but alive.

While recovering in the hospital, Mauricio told reporters about the fortune in currency he had discovered deep in the cavern. The next day newspapers all across the American Southwest carried the story of this incredible treasure found in the remote cavern near Van Horn. In a very short time, dozens of people and representatives from a variety of agencies were submitting claims for the ownership of the land on which the cave was located as well as the treasure within. For at least one week, the sheriff's department posted guards at the tiny entrance of the cavern in order to keep outsiders from entering it. Finally a judge relieved the sheriff's department from the obligation of guarding the hole and placed the responsibility with Mauricio and his friends.

At the time, the property was owned by a land development company. Two days after the find, a representative of the company flew out to West Texas from California to make a claim on the treasure. At a hearing, the judge declared that Mauricio and the development company should share the costs and effort related to bringing the treasure to the surface. Once recovered, said the judge, it was to be turned over to the court for safekeeping until a proper determination of ownership could be made by a trial court. In the meantime a representative from the Texas Parks and Wildlife Department arrived with documents stating that, by law, the state of Texas would own any and all items "of historical significance" recovered from the cave.

One week later it was decided that a team of cave explorers and other experts from the University of Texas at El Paso should be allowed to enter the cave. A representative of the president of the university even traveled to the site to oversee the operation.

Members of the exploration team descended the full 400 feet to the bottom of the deep chamber. Here, they found Mauricio's flashlight along with a pair of bolt cutters. On arriving back at the surface, the members of the team claimed they saw no lateral passageways in the cavern whatsoever. Mauricio, who was standing nearby and overheard the comment, stepped forward and pleaded with the cavers to go back down and conduct a more thorough search. The passageways, he maintained, were indeed there. The cavers, however, flatly refused, packed up their gear, and departed.

Several days later Mauricio's lawyer stated that another attempt would be made to enter the cavern in order to retrieve the treasure. In the meantime, however, he said he would obtain a court order to place a locked gate over the entrance to prevent anyone without the proper authority from entering the cave.

As a result of ongoing court delays as well as a variety of legal complications, no one has been allowed to reenter the cave to retrieve the cache of currency Mauricio claims is still there. While it has never been verified, a rumor surfaced some years ago that someone had opened the locked gate, entered the cavern, and, using mortar and rock, sealed up the entrance to the lateral passage that allegedly contained the hidden treasure!

Yet another rumor has made the rounds in recent years, and it suggests that a group of unnamed individuals is financing the excavation of an adjacent shaft that would enter the deep cavern at an angle. Since the rumor became known, a number of concerned individuals have taken turns keeping watch on the location to determine who is behind the operation, if in fact it is true.

Clearly, if someone is willing to invest the requisite time, money, and energy into getting into the cavern, they must have some knowledge relative to the validity of the existence of the treasure believed to lie within.

Though several investigative reporters have attempted to discern the origin of the suitcases and cloth bags filled with what must be hundreds of thousands of dollars' worth of currency in the cavern, nothing of substance has been learned. Where the money came from remains a mystery, and why and how it got into the remote cavern in the West Texas desert remains an even greater mystery.

Eagle Mountains Ledge of Gold

Somewhere in the Eagle Mountains of West Texas lies an exposed ledge of gold that has tantalized treasure hunters and adventurers for well over a century. Foremost among those who have searched for, found, and subsequently lost this mysterious ledge of gold was the noted soldier, politician, and entrepreneur Jacob Snively.

Jacob Snively was no stranger to precious metals, mines, and mining. Using some documents he accidentally discovered, Snively eventually located a long-forgotten Spanish silver mine deep in South Texas. Snively was also one of the first to locate and mine the rich placer deposits on the Gila River in Arizona during the 1850s. In addition, he found the famous placer mines in the Pinos Altos region of New Mexico in 1860, harvested huge gold nuggets from the wilderness near the Arizona-New Mexico border, and was one of the organizing forces behind the immensely rich Castle Dome mining enterprise centered in La Paz, Bolivia.

While prospecting for precious metals in southwest Texas, Snively encountered a soldier lost, alone, and dying. The soldier had become separated from his company several weeks earlier following a skirmish with the Apache Indians not far from the Rio Grande. Suffering from a serious wound, he survived by foraging for edible roots and berries and occasionally eating a snake or a lizard. Knowing the soldier did not have long to live, Snively saw to it that he was made as comfortable as possible.

The soldier was provided with the miner's personal tent, and one evening as Snively was spoon-feeding the dying man, he was told a most remarkable story about an incredible deposit of gold ore.

When he finished eating, the soldier asked Snively to hand him his jacket. Shakily, the soldier reached into one of the pockets and withdrew a handful of gold nuggets, which he handed to the miner. Snively examined them closely and found them to be quite pure. The soldier told Snively that his saddlebags were also filled to the top with similar nuggets.

The soldier went on to explain to Snively that, while traveling through the Eagle Mountains located in Texas's Hudspeth County, he paused to water himself and his horse at a spring in the foothills of the small range. While resting beside the spring, the soldier spied several small objects glittering at the bottom of the clear, shallow pool of water. Reaching into the water and retrieving them, he was surprised to discover they were gold nuggets! Believing the nuggets had made their way into the pool from some location higher up, he proceeded to explore the adjacent slope. After about two hours of searching, he encountered what he referred to as an exposed "ledge of gold." The ore was so pure, he stated, that it could be easily pried out with only a pocketknife.

The soldier gave Snively directions to the spring. The next morning as Snively was preparing breakfast, he paused to check on his guest only to discover the soldier had passed away sometime during the night.

Snively was vaguely aware of the location of the golden ledge in the Eagle Mountains as described by the soldier, having passed through that particular region on at least one previous expedition. Snively always remained optimistic about the possibility that precious metals could be found in that area but was never presented with the opportunity to search for them. One reason for not exploring the area earlier was the fact that it was occupied by bands of hostile Apaches. These Indians were constantly at war with soldiers, settlers, miners, prospectors, and any and all who would dare pass through their homelands.

Snively determined that in order to search for, locate, and mine the gold from the mysterious ledge in the Eagle Mountains, he would need to assemble a large force of thirty to forty men who would not only dig for the gold but also be willing to fight off potential attacks from the Apache Indians who lived in the region.

Returning to San Antonio, Snively began making plans to organize a forty-man expedition to return to West Texas and the Eagle Mountains. He immediately contacted his old friend and former fellow soldier Colonel W.C. Dalrymple. On hearing Snively's plan explained, Dalrymple quickly discouraged the idea, claiming that ten experienced fighting men could handle the Apaches and then each participant would receive a larger share of the gold. Snively ultimately agreed and during the next few days gathered together an impressive group of marksmen, hunters, adventurers, and ex-soldiers, including Mose Carson, the brother of Kit Carson. They departed in January 1867, leading a pack train of over one hundred horses and mules, and headed for the Eagle Mountains and potential riches.

During their trip west, the Snively party was joined by other men they met along the way, eventually swelling their number to seventeen. One night while camping just west of the Concho River, the group was attacked by a band of approximately 200 Comanche and Kiowa Indians. Fighting was fierce and at times involved hand-to-hand combat. Those of the party who were not killed outright by the Indians suffered a variety of wounds.

During the fighting, the Snively party managed a somewhat disorganized retreat back toward the Concho River, hoping the Indians would cease to attack. It was not to be, and the men eventually took shelter in a dry wash, which afforded them a modicum of protection and from which they could fire at the Indians.

The fighting lasted well into the late afternoon and only lightened up somewhat with sundown. The Snively party huddled in the wash until midnight and then, as a dust storm began kicking up, quietly snuck away in the night, continuing eastward toward the Concho River and, hopefully, reinforcements. Three days later the survivors, convinced they had outdistanced the Indians, decided to separate and return to their homes. Just before disbanding, they agreed to rendezvous four months hence at Ben Ficklin's ranch, later to become Fort Concho, located not far from their previous camp near the Concho River. The next time, promised Snively, they would proceed to the Eagle Mountains with a larger force and make another attempt at finding the ledge of gold.

In all, approximately fifty men eventually gathered at Ficklin's ranch. After inventorying pack animals, mining equipment, and other supplies, the large party undertook the long journey to the west. Along the way they encountered Indians on several occasions, but superior numbers and armament kept the potential raiders at bay. On June 21 the party forded the Pecos River at Horsehead Crossing and entered the severe environment of the West Texas Chihuahuan Desert, constantly on the alert for warring Apaches.

Without serious mishap they arrived at the spring where the dying soldier had told Snively about the gold nuggets he found lying in abundance. While the men were setting up camp, Snively and several others examined the spring and actually found and retrieved a handful of small nuggets. It was clear to the men the nuggets had been washed down into the pool from some higher elevation along the adjacent slope. In the morning, Snively told them, they would inspect the slope, locate the outcrop, and begin extracting the rich gold he was certain would be found there.

The next morning Snively and over a dozen men crawled and climbed over the slope above the spring. All day long they searched for the ledge of gold described by the soldier but could not find it. The search was continued the following day and for many days thereafter, but after two weeks not a single sign of the ledge of gold was found.

Other parts of the range were also explored and searched, other springs were visited and examined, and still nothing could be found. After six weeks in the Eagle Mountains with no results, Snively's charges were growing irritable and threatening. In spite of Snively's assurances that the gold would be found eventually, the group finally disbanded. Snively himself proceeded on to Arizona where he had a few mining interests. While there, he began making preparations for a return expedition to the Eagle Mountains, but he, along with some fellow miners, was killed by Indians the following year near Wickenburg.

Several years later the ledge of gold was found, this time by a man named John Ryland. Ryland was traveling from El Paso to San Antonio and decided to take what he believed would be a short cut through the Eagle Mountains. As he rode through the range, he noticed his canteen was nearly empty so he began looking for a spring. Two hours later he found one, and while filling his canteen he noted a mineral glimmer in the small pool of water. Carefully sieving the sands and gravels through his fingers, Ryland was surprised and delighted to discover a number of small gold nuggets in the pool. Presuming, as others did before him, that the origin of the gold was likely somewhere along the slope rising just above the spring, he ascended it and examined several likely outcrops. Finally he found a thick seam of bright, shining quartz that was laced with the purest gold he had ever beheld. He dug out several chunks of the gold-laden quartz, stuffed them into his shirt, returned to the spring and his horse, and continued on to San Antonio. Along the way, Ryland began making plans to return to the spring and mine the rich ledge of gold located above it.

At the first opportunity in San Antonio, Ryland organized a group of fifteen men, purchased equipment and supplies, and returned to the Eagle Mountains. Certain that he had relocated the exact same spring from which he filled his canteen during the previous trip, he searched the adjacent slope for the ledge of gold, but like Snively, Ryland was unable to find it. The disappointed members of his party grew hostile and threatened to kill Ryland but subsequently decided to simply abandon him in the mountain range. While the rest of the party returned to San Antonio, Ryland went back to El Paso and was never heard from again.

Yet one other man found the ledge of gold in the Eagle Mountains. Like the others before him, he was convinced he could relocate it and was determined to grow rich from the ore it yielded. But, as with the others who came before him, fate intervened and it was not to be.

In 1875 W.W. Williams was working as a member of a surveying team that was laying out the path for a San Antonio to El Paso railroad line along a portion of the Rio Grande. Their work was quite dangerous, for hostile Apaches still roamed these remote lands among the Quitman and Eagle Mountains.

While working near the Eagle Mountains, Williams and his crew ran low on water. One of the guides, an Ysleta Indian, told Williams he knew of a spring located near the head of one of the nearby canyons in the range. Later that afternoon Williams and his surveying crew packed up their gear and traveled into the canyon, intent on filling several barrels with water and camping that night near the spring.

The next morning two of the surveyors found some gold nuggets in the tiny stream that flowed from the spring, and Williams found more gold nuggets in the spring itself. Curious, the surveyors climbed the slope in hopes of finding the source of the gold and apparently found the same vein of quartz Ryland did, for Williams's description matches his exactly. Williams dug out several large pieces of the gold and made note of pertinent landmarks. In his journal Williams recorded that from where he stood near the ledge of rich gold, he could see the Rio Grande to the southwest as well as the flag flying on the pole at nearby Fort Quitman. To the north, he wrote, he could see two peaks that gleamed white in the sun. Williams also wrote that in all his life, he had "never seen such a ledge so rich in gold...."

Williams and his crew returned to work, but the next day a war party of Apaches was spotted and the men hastily packed their equipment and fled from the area. In the days that followed, so many Indians were spotted in the area that the survey foreman decided it would be prudent to abandon the work for the time being and return when the Apache threat was over. Williams went home to San Antonio. There, he took the gold he had extracted from the ledge to an assayer and was informed that it was quite rich. Like those who went before him, Williams organized a group of men to travel back to the Eagle Mountains to mine the gold. Before they ever made it to the range, however, Williams's party was attacked by Indians, two of its members killed, and several more wounded. Dejected, Williams returned to San Antonio.

Years later Williams made another attempt to find the ledge of gold, but on entering the Eagle Mountains he became disoriented, found the landmarks confusing, and was never able to relocate the spring. Several more expeditions into the range to find the ledge of gold likewise ended in failure, and Williams eventually moved to Colorado, never to return to Texas.

It seems quite likely that Snively, Ryland, and Williams all found the same ledge of gold. Why these three men were never able to relocate this mysterious ledge remains a mystery, but such things are not uncommon in the desert where the sun and shadows play tricks on visitors.

Apache Gold in the Eagle Mountains

The lost ledge of gold is not the only tale of lost wealth associated with the Eagle Mountains of West Texas. Deep in the heart of the range lies hidden a cave, one that is believed to be filled with gold bars stolen from Mexican pack trains by Apache Indians.

During the middle of the nineteenth century, more and more Americans journeyed into the western states of the country all bent on making their way and their fortune in the promising lands opening up to settlement. They were trappers, traders, drummers, miners, preachers, and entrepreneurs,.

In response to the increasing numbers of travelers, railroad companies and stage lines opened up new routes to accommodate the ever-increasing customers. One such stage line wound across the plains and deserts from San Antonio to El Paso. Along the way, it stopped at a number of previously established stations to change horses and provide meals and sometimes beds for the customers.

One such station was located at Eagle Springs, Texas, in a location in the foothills on the north side of the Eagle

Mountain range approximately fifteen miles southeast of Van Horn.

One of the stagecoach drivers was a young man named Joseph Peacock. Though only nineteen years of age, Peacock had already survived several shootouts with Indians and outlaws. He was also a skilled and dependable driver and quickly became a favorite among the station managers and employees. The stagecoach company desperately needed men of Peacock's caliber to negotiate the coaches through this rugged part of West Texas, a region constantly under the threat attack by of hostile Apache Indians. During the few weeks that Peacock had been driving the stage to El Paso and back, Apache chief Victorio, who had a reputation for despising all white men, was seen roaming this area.

One afternoon just as Peacock pulled into the Eagle Springs stagecoach station, approximately twenty warriors charged from hiding out of the nearby mountains, firing rifles, hurling lances, and shooting arrows into employees and travelers alike. Within a few short minutes everyone was killed save for Peacock, who was struck in the thigh by an arrow and taken captive.

After tying the wounded driver to a stout Indian pony, Victorio led his band of warriors far to the south into Mexico and the Tres Castillos Mountains, a long-time Apache stronghold. Here they were safe from pursuit by the U.S. Army, and they could rest up before launching another raid on the hated whites to the north. As the tightly bound Peacock rode along, he felt certain he was being saved for some form of horrible torture at the hands of the Indians. He had heard tales of what they did to white men, and he shook with fright at the notion that it would likely happen to him.

On arriving at the Tres Castillos Mountain hideout, Peacock was untied and his wound was treated by a young woman named Juanita. Though it has never been determined with certainty, some researchers are convinced Juanita was a daughter of Chief Victorio. When Peacock's wound began to heal, he was forced to work around the Indian camp, cutting and gathering firewood. From time to time he was beaten by warriors and women alike, but as long as he lived he harbored hope he might find a way to escape.

In the meantime Juanita began spending more and more time with the captive, sometimes sneaking out of her wickiup at night to be with him. On more than one occasion, Victorio threatened to kill Peacock, but Juanita successfully pleaded that his life be spared.

One night, in a display of uncharacteristic boldness for an Apache woman, Juanita told Peacock she would marry him. Though Peacock was not in love with Juanita, he did not want to refuse the only person responsible for keeping him alive in the Apache camp. He made up the story that his mother back in Texas was ill and that he was very concerned about her. Once he was certain she was well, he told Juanita, he would marry her.

Juanita accepted Peacock's explanation for a time but then began pressing him to marry her even more. During her courtship, she told Peacock that if he would consent to wed her, she would tell him where Victorio hid all of the gold he had stolen from Mexican pack trains. Peacock had seen a number of gold ingots in the camp on several occasions, and he learned that the Indians used them to trade for rifles and ammunition from Mexican and American gunrunners. He also watched as Apache artisans fashioned bracelets and amulets from the soft metal. The gold ingots seemed to be plentiful, and Peacock often wondered to himself about where it might be cached.

When Peacock manifested interest in the Apache's gold, Juanita explained to him that hundreds of bars were stored in a small cave in the Eagle Mountains not far from the stagecoach station, and that there was a well-marked trail that led from it to the tiny settlement of Indian Hot Springs on the Rio Grande. Juanita told her lover that as a child she had visited the cave many times with her father and had often played with the gold bars. In addition to the gold ingots, she claimed, there were also dozens of buckskin sacks containing gold coins and gold nuggets. She explained to Peacock that it would take at least fifty burros to transport all of the gold from the cave.

Juanita also informed Peacock that a small party of Apaches recently traveled to the cave to retrieve some of the gold. Along the way they were spotted by a platoon of U.S. Cavalry, which gave chase. A brief skirmish ensued in which two of the soldiers were killed and one of the Apaches seriously wounded. The soldiers retreated, but on hearing the report of the incident, Victorio grew convinced they would return with a much larger force. Victorio feared the soldiers would discover the huge cache of gold in the small cave in the Eagle Mountains, so he sent another party of Indians into the range with orders to conceal the entrance. It is believed they did so by stacking rocks in front of the small opening and then generating a small landslide, thus making the site look much like the rest of the mountain.

With Juanita's directions to the treasure cave, Peacock began making plans to escape from the mountain stronghold. At night while he slept, he dreamed of the incredible riches that awaited him in the Eagle Mountains.

Weeks passed, weeks which afforded Peacock no opportunity to escape since he was watched closely all hours of the day and night. Then early one morning as he was gathering firewood, he noted that Victorio, after assembling most of the warriors, rode south into the heart of the Mexican state of Chihuahua to conduct raids and take horses. The village, noted Peacock, now consisted mostly of women, children, and old men.

That evening Peacock persuaded Juanita that it was now a good time for him to leave. He promised her he would go see to his ailing mother and return for her as soon as he could. She brought him a sturdy Apache pony, a saddlebag filled with jerked venison and tortillas, and a deer gut filled with water. Bidding Juanita goodbye, Peacock quietly rode out of the Indian village, down the mountain, and toward the north and the safety of the United States.

Using the directions provided by Juanita, Peacock rode directly to Eagle Springs where he set up camp. Here, he rested for approximately one week before undertaking the search for Victorio's cave of gold. Several times Peacock rode and walked along the Indian trail in the area described by Juanita, but he had difficulty interpreting the landmarks. Though a number of locations along the trail matched her descriptions, many of them did not. Following several more days of searching, Peacock became confused. After spending an entire month looking for the cave and finding nothing, Peacock wondered if Juanita had given him only partial directions in the hope he would return to her.

Soon Peacock exhausted his provisions and was forced to find work so he could purchase more. Eventually he regained his job as driver for the stage company, and in his spare time he continued his search for Victorio's gold.

Years passed, and Peacock managed to save enough money to purchase a ranch not far from the Eagle Mountains. From here, he could pursue his search for the amazing wealth in gold he knew lay hidden in the small cave somewhere tantalizingly close to the old Indian trail.

During the year 1880, a Texas Ranger command under the leadership of Baylor and North was patrolling along the Rio Grande and in the Eagle Mountains. Their assignment was to keep an eye out for Indians and engage them when necessary. They had recently received information that the Apaches had just been attacked in their mountain stronghold by Mexican soldiers led by General Terrazas, who intended to slaughter them all and remove the Apache menace from his country once and for all. Should any of the Apaches escape and attempt to flee toward the north, the Texas Rangers would be prepared to intercept them. Tired of waiting, the rangers crossed the river and traveled into Mexico toward the Tres Castillos Mountains. Riding with the rangers was Joe Peacock, who had signed on with them only weeks earlier.

By the time the rangers arrived at the Indian stronghold, all of the Apaches had been killed or captured, with only a handful managing to escape. Victorio was dead. Though Terrazas claimed his soldiers killed him, the Apaches maintained the chief took his own life rather than be captured.

Upon learning that approximately one dozen Apaches managed to elude the soldiers and were fleeing toward the Rio Grande, the rangers set off in pursuit. On crossing the river, Baylor noted that the tracks of the fleeing Indians led toward the Eagle Mountains.

Three days later the rangers encountered the Apaches in the range, forcing them to take shelter among some boulders. For nearly twenty-four hours the two groups exchanged gunfire, but eventually the Indians ran out of ammunition as well as food and water. Unaware the Indians had exhausted their firepower, the rangers held their positions, waiting for a chance to attack. During the night with only the light from a quarter moon, the Apaches quietly crept from the protection of the rocks and made their way past the rangers and out of the Eagle Mountains. All night long they hiked and rode until finally reaching the Sierra Diablo range several miles to the north. They were climbing among the foothills and making their way to the higher elevations of the mountains when they were overtaken by the rangers. A brief battle ensued, but the outnumbered and outgunned Indians fell victim to the rangers' attack. All were killed.

Following the skirmish, Peacock searched the faces of the dead women to see if Juanita was among them, but she was not. He presumed she had been killed or captured at Tres Castillos. In any event, he never saw her again.

More years passed, and Joe Peacock, now an old man, continued to raise cattle not far from the Eagle Mountains. Though it was increasingly difficult for him to get around, he occasionally found the time to travel into the range to search for the elusive cave containing Chief Victorio's gold.

During one of his trips to the Eagle Mountains, Peacock met a young man named Race Compton. Compton was traveling on foot and decided to linger in the Eagle Mountains for a while and do some prospecting. Compton and Peacock spent a great deal of time together, sharing conversations around the campfire and relating prospecting and mining stories. Eventually Joe Peacock told Compton the story of Victorio's cave of gold. For the next fifteen years, the two men searched together for the cave, but it continued to elude them.

Joe Peacock died in 1915. Race Compton continued to live on Peacock's ranch while he searched for the gold-filled cave. One day following a heavy rainstorm, Compton noticed a place located somewhat above the trail where the runoff had washed away some rocks from one particular slope. Climbing up to the site, Compton was surprised to discover the outline of the entrance to a cave, an entrance that had been stuffed with heavy boulders. He labored for several hours in an attempt to remove the boulders, but he lacked the strength to budge any but the smallest of them. He decided he would hitchhike to El Paso, some 100 miles to the northwest, and purchase enough dynamite to blow the heavy boulders out of the opening. On his way to El Paso, Compton revealed his discovery to a neighboring rancher, but he provided no details with regard to the location.

On his return trip, Compton caught a ride as far as the small town of Sierra Blanca. After being dropped off, he began walking eastward along the highway toward the dirt road some miles distant that would take him into the Eagle Mountains. On his back he carried a knapsack filled with sticks of dynamite he intended to use to open the cave.

Race Compton never made it back to Victorio's treasure cave. Only a few miles out of Sierra Blanca, he apparently suffered a heart attack and died. When he was found, he was lying on the roadside, his head resting against the dynamite-filled knapsack.

Race Compton apparently left no notes relative to the location of the treasure cave. His few meager belongings were searched, but nothing was ever found save for old, ragged clothes, a few battered cooking utensils, and a pistol that had not been fired in years.

Today the Eagle Mountains, though easily seen from Interstate 10 in West Texas, are seldom visited save for a few deer hunters now and then and the occasional rancher searching for cattle that may have strayed into one of the canyons. Near the old stagecoach station, the state of Texas has placed a historical marker containing just a few words.

The gold of Victorio, presumably millions of dollars' worth, still lies concealed in the small cave somewhere in the range. Occasionally a treasure hunter will enter the region in search of the cave, but this harsh and arid environment discourages all but the most rugged and hardy.

Monahans Sandhills Treasure

Lying near the point where corners of Winkler, Ector, Crane, and Ward Counties all come together in West Texas can be found an expanse of sand dunes rising out of the otherwise gravelly and creosote bush-littered floor of the Chihuahuan Desert. From an airplane, this expanse of sand appears much like an island. So unique and spectacular are these dunes that they have become

Monahans Sandhills State Park, a travel destination that receives thousands of visitors each year. Little do the visitors to this park know that, as they hike and play across the dunes, somewhere just a few inches below the surface near a spring lies buried the remains of some forty wagons along with 477 twenty-pound gold ingots, a lost treasure that would be worth nearly $30 million if found today.

In 1873 preparations were being made for a wagon train journey from Yuma, Arizona, to St. Louis, Missouri. The party of travelers consisted of several dozen families of Dutch descent, all of whom journeyed to California years earlier to seek their fortunes in the gold fields of that rich and prospering state.

They did, indeed, find their fortunes, for each of the families were successful in gleaning riches from the gold-laden rock matrices of the California Rockies as well as the numerous streams that coursed through the area. The gold was mined, panned, smelted, and accumulated until it was decided that enough had been earned to place toward the establishment of a Dutch farming community near the Missouri-Illinois border.

The finest wagons, horses, and oxen were purchased for this long trip, and into several of the wagons were loaded the gold bars along with all of the families' possessions. It was late September, and the Dutch families were growing more eager with each passing day to begin the long trek to the east.

Travel was often difficult: Floods, storms, drought, hostile Indians, and sometimes outlaws slowed the pace of the travelers, and it was nearly three months later that the wagon train finally forded Horsehead Crossing on the Pecos River in Texas. Only minutes after the crossing, the leader of the party, a man named Daniel Flake, spotted several mounted Indians watching the wagon train from a nearby rise. Flake rode toward the Indians in an effort to determine if they were friendly or not. When he was about fifty yards from their position, however, the Indians turned and galloped away along an adjacent ravine. Flake had gotten close enough to identify them as Comanches, one of the most feared tribes in North America and one that continually attacked and killed settlers and travelers they chanced upon in their territory. Following the brief encounter with the Indians, Flake doubled the guard.

For the next few days, Comanches were often observed watching the wagon train from a distance, but when Flake or other riders attempted to approach them, they simply rode away, gradually fading into the landscape.

Flake was harboring a bad feeling about the continuous presence of the Comanches and confessed to one of the stock handlers that he had recurring premonitions of disaster.

Travel remained slow and tedious, and it was a few days later when the wagon train finally passed through the tiny settlement of Monahans, Texas, a community that consisted of little more than ragged tents and brush shelters. Here, a scout rode up to Flake and reported that just beyond the settlement and among a huge cluster of sand dunes he found a freshwater spring. It was there Flake decided to camp for the night, and he and the scout led the party in that direction. As they traveled, Flake noticed some Indians watching them from horseback from a rise in the ground some 200 yards away. Flake counted over 150 of them. He also noted that they were all carrying bows and lances at the ready.

Around two hours later the wagon train entered the expanse of sand dunes. The hills of shiny, light-colored grains of sand extended to the east and north as far as Flake could see, and some of the dunes were dozens of feet high. The tall dunes, along with the soft, yielding sand, made travel quite difficult. Horses and oxen found that moving through the soft and yielding sand was far more exhausting than the hard-packed desert floor they had grown used to. The wheels of the heavy wagons sometimes sank into the soft sand clear up to the axles, necessitating the efforts of men and additional draft animals to pull them out. As the Dutchmen fought the challenge of these sandy obstacles on their way to the freshwater spring, the Comanches were slowly walking their stout ponies closer and closer to the wagon train. Presently, they sat nearly motionless on horseback only forty or fifty yards away and watched as the Dutchmen struggled with the wagons.

It was well after sundown when the party finally reached the spring. Flake ordered the men to arrange the wagons in a protective circle should the Indians decide to attack. At the same time he instructed the women and children to fill the water barrels from the shallow pool at the spring.

It was a bright waxing moon that night, and as Flake and several of the Dutchmen stood guard, they could see the Comanche riders slowly and deliberately circling the campsite. Throughout the night the Indians rode in an ever-widening circle until just before dawn when they quietly disappeared.

During the night three of the Dutch elders approached Flake and voiced their concerns about the possibility that the Indians knew about the gold they transported and wanted to steal it. They suggested a plan to the wagon train leader wherein the ingots would be buried at this particular site and the wagon train would continue on to Missouri. When the time was right and the Indian menace removed from this region, appointed members of the group would return and retrieve the gold. Despite Flake's assurances that the Comanches cared little or nothing for gold, the elders insisted. When it was clear they were not going to retreat from their proposal, Flake agreed to let them unload the gold ingots from the wagons and bury them in several large excavations just inside the circle of wagons.

As dawn slowly broke over the eastern horizon the next morning, Flake, who remained awake and worried the entire night, was pacing the margins of the campsite, looking for signs of the Indians. Within minutes he spotted one—a lone rider carrying a lance sitting astride his horse atop a dune some forty or fifty yards away. As Flake watched the Indian, the rest of the camp began to stir as the travelers awakened and began thinking about breakfast. Fires were being started and horses and oxen tended to.

As the Dutch bustled about, the lone Comanche slowly raised his lance, and within seconds he was joined by what Flake estimated to be another 200 Indians, all armed with similar lances, bows, arrows, war clubs, and even a few rifles.

As Flake was considering approaching the Indians to parley, the relative quiet of the early morning was shattered by a piercing war cry. Seconds later the Comanches were thundering down the dunes toward the circle of wagons, their war cries ringing through the air and mixing with the screams of the frightened travelers.

As the women and children ran for the shelter of their wagons and as the Dutchmen searched for and loaded their rifles, the Comanches were upon them, dashing about the camp on horseback within the circle of wagons. Shooting, slashing, and stabbing their way through the panicked travelers, the Indians fought fiercely for thirty minutes until all members of the wagon train—men, women, and children—were dead.

When the last of the migrants were scalped, the Comanches began looting the wagons, taking bolts of cloth, cooking pots, clothes, and other items deemed important. Unwanted goods were thrown to the ground. A group of warriors cut the harnesses and reins from the stock and began herding them away toward the north.

By noon the wagons had all been looted and set afire, and the last of the Indians had ridden away, joining the main band that traveled northward. The afternoon desert winds were already stirring the sand grains on the desert floor, blowing them up against the destroyed wagons and the scattered and mutilated bodies of more than a hundred Dutch. Buzzards that had been circling high in the sky were now descending in ever-tightening spirals toward the corpses lying on the blood-soaked sand.

The gold ingots, their total weight estimated to be as much as 10,000 pounds, lay just below the shifting sands in several individual caches, unseen by the Indians. They would remain there for a long, long time.

With the passage of decades, the ever-shifting winds eventually closed in upon and covered the remains of the wagon train. Travelers to the freshwater spring occasionally told of finding what appeared to be an old wagon plank or a piece of harness, but none were aware of the tragedy that had occurred here during the early part of 1874.

Other than availing themselves of the cool waters of the spring, travelers in this region had little reason to pass through these sandhills. The first formal expedition to enter this area was led by one Arthur Hayes, a retired U.S. Army colonel and a prominent judge in this part of West Texas. Hayes undertook such expeditions from time to time to evaluate the surrounding region and assess its possibility for the grazing of livestock.

The Hayes expedition, led by a scout named Robert Brown, entered the Monahans sandhills and set up camp at the freshwater spring, now called Willow Spring. While encamped here, Hayes and members of his party found several charred pieces of milled timber similar to the kind used to construct wagons. They also found some wagon wheels and several fittings lying about partially buried in the sand. On closer examination, they found several dozen human skeletons. Some of the skulls had been obviously smashed by heavy instruments, and stone arrowheads were occasionally found embedded in several of the limb bones. After piecing together the evidence encountered at Willow Spring, Hayes tried for several years to solve the mystery of the destroyed wagon train and its members, but he passed away before learning the truth. In Hayes' wake, other researchers analyzed the information and eventually concluded the site contained the remains of a long lost wagon train that left Yuma,

Arizona, in 1873 and never showed up at its intended destination in Missouri.

Several years later one of the researchers found an old newspaper clipping describing the formation of the party of Dutch travelers and told of the precious cargo of 477 gold ingots. Excited by the prospect of finding the treasure, the researcher returned to the site near Willow Spring only to find that it had been once again covered over by the shifting sands of the desert. He was unable to find anything.

Since the connection was finally made between the massacre of the Dutch travelers near Willow Spring and the huge amount of gold bars they were carrying, hundreds of hopeful treasure hunters have arrived at Monahans Sandhills State Park over the years to try to find this impressive treasure. Now and then another piece of wagon fitting or leather harness is found, discoveries that convince the searchers they are in the right place. To date, however, the gold has never been found.

The Mysterious Treasures of Castle Gap

Castle Gap is a 400-foot-deep pass through a limestone mesa in western Upton County. This location, a few miles southeast of the town of Monahans, is the setting for a number of rather mysterious lost treasures. Enough evidence exists to suggest strongly that these treasure do, in fact, exist; however, their exact location in this rugged, arid, and somewhat forbidding West Texas environment remain unknown.

One of Castle Gap's most famous treasures is known as the Catholic Cross Cache. This particular hoard is believed to consist of fifteen burro-loads of gold, silver, and jewels, the vast majority of which was stolen from at least three Catholic churches in Mexico City, probably sometime during the year 1780. Much of the gold and silver was in the form of chalices, vases, crosses, and other church-related items. The thieves are believed to be three priests whose names have been lost to history, and most who have studied this particular treasure are convinced that their motive for robbing the church was intense dissatisfaction with religious policies.

Regardless of the reason and the exact identity of the perpetrators, it is known that the thieves avoided obvious coastal destinations during their escape and, with the help of approximately a dozen Indian slaves, led their treasure-laden pack train out of Mexico City and far to the north into the Chihuahuan Desert, eventually crossing the Rio Grande and making their way through West Texas. From here, it is believed they intended to travel east to New Orleans where they intended to book passage on a ship and carry the treasure back to church headquarters in the Vatican.

The route through the Texas portion of the Chihuahuan Desert proved to be quite difficult: Lack of water and game led to prolonged periods of starvation and thirst, and those who did not succumb to lack of food and water fell prey to Indian attacks. By the time the party crossed the Pecos River at Horsehead Crossing and was passing through what is now called Castle Gap, the weary and hungry group was decimated. The prospect of herding the fifteen heavily packed burros to New Orleans seemed a daunting task for the handful that remained. The priests decided it would be best to bury the church treasure, travel to New Orleans, and return with a larger and better-equipped force to retrieve it.

The treasure, as the legend goes, was buried somewhere at Castle Gap, and the bodies of two of the Indian slaves, recent victims of a Comanche attack, were buried along with the church wealth. One version of the tale has the treasure buried at or near the top of V Mountain. Another version states the treasure was buried at or near the foot of the mountain. Following the caching of the treasure, the priests inscribed various directional markings on exposed rock surfaces throughout the area, including a huge cross, as indicators to aid them in finding the treasure on their return.

The survivors continued northeastward, eventually arriving at present-day Big Spring. Here, according to some scholars, they carved an inscription on a large rock that read, "Look for huge cross, 77. Then look for V-shaped mountain." The inscription, rather than shed pertinent information on the location of the treasure cached, merely clouded the issue. Many believe that "77" referred to the number of miles the treasure lay from the inscribed rock and have pointed out that Castle Gap is located seventy-seven miles to the southwest and that V Mountain is a prominent landmark. The Spanish, however, did not measure distances in miles, but rather in leagues during that time.

The small party consisting of the three priests and a handful of surviving Indians continued on but apparently didn't survive the journey, for they were never heard from again.

A rather curious related event occurred a few decades later. Sometime during the early 1800s—no one knows when—an aged woman, along with two of her grandchildren arrived at Castle Gap in a wagon pulled by two mules. According to those who spoke with her, she claimed to have traveled all the way from El Paso and that she was one of the Indians who was part of the original party that carried the church treasure from Mexico City to this area and buried it. Leaving her grandchildren with a relative in Monahans, she drove the wagon to the base of V Mountain (called Red Mountain then by some) where she set up camp. Now and again she was seen by ranchers or travelers climbing around on the mountain. Several days later she returned to Monahans, retrieved her grandchildren, and was overheard telling one of them the treasure was still there.

The old woman died before she was able to retrieve any of the buried wealth. Numerous Spanish inscriptions have been found on exposed rocks at V Mountain, but to date no one has ever deciphered them.

In 1968 a college student was exploring around this region on foot and searching for arrowheads when he made a most amazing discovery. Runoff from recent rains had washed away a portion of a low-angled slope exposing some bones. On investigation, the student discovered the remains of two Indians—he deduced they were Indians by the trappings found along with the bones. Manifesting a level of respect for the dead, he reburied the skeletons and continued on. In 1974 when he learned the story of the Catholic Cross Cache and realized that the two skeletons were likely the ones placed atop the buried church treasure, he returned to Castle Gap to try to relocate the site but was never able to find it.

The treasure, estimated to be worth several million dollars, apparently still lies where the priests originally buried it back in the 1780s.

Another famous treasure believed to be hidden in the Castle Gap area is the one called Maximilian's Treasure, an incredible fortune consisting of fifteen wagon-loads of gold and silver packed into wooden barrels.

In 1864 Ferdinand Maximilian, an Austrian prince, had himself appointed Emperor of Mexico. His decision to rule was backed up by a large army supplied by his cousin, France's Napoleon III.

Maximilian ruled in a manner that made him quite unpopular with the citizens of Mexico, and when Napoleon recalled his army three years later, the Emperor realized he had little in the way of military strength to help him hold onto the throne. Convinced that he would be assassinated if he remained, Maximilian, in the manner of many despotic rulers before him, sacked the country's treasury, had all of the gold and silver packed into wooden barrels and loaded onto wagons, and then fled the country.

Maximilian appointed a trusted aide and fellow Austrian countryman to oversee the shipment of the wealth out of Mexico and into Texas. From there, he intended to have it shipped to his Austrian homeland.

By the time the treasure caravan reached Ciudad Chihuahua several weeks later, one of the scouts reported to the aide that the trail ahead was extremely dangerous because of raiding Indians. In response, the aide added a number of armed escorts to the caravan. With the teamsters driving and maintaining the wagons, the initial escort, and the added men, the party grew to well over one hundred. Though they were subjected to at least three Indian attacks during the journey northward, the well-armed force of Mexicans was able to repel the invaders. Eventually the party crossed the Rio Grande at the Mexican village of Ojinaga. Here, the majority of the escort decided to turn back, leaving the precious cargo relatively unguarded. In desperation, the aide employed six men he met just across the river in the Texas town of Presidio. The six told the aide they were ex-Confederate soldiers, that they were from Missouri, and that they had ridden with Quantrill in Kansas.

The trip from Presidio to Horsehead Crossing on the Pecos was uneventful. After crossing the river, the party set up camp to rest horses and men for a few days. While encamped in a grove of trees on the bank of the river, one of the Missourians grew curious as to what was being transported and why it needed to be guarded. Late one night while everyone was asleep, he untied the canvas top to one of the wagons, broke into a barrel, and discovered the gold and silver. After replacing the tarp, he informed his companions, and for the rest of the night the ex-Rebels plotted to steal the treasure shipment.

Several days later as the wagons creaked along the rough trail that wound through Castle Gap, the Missourians attacked. With surprise on their side, they quickly subdued the rest of the group, killing them all. The remainder of the day was spent excavating a wide, deep hole just off the trail. When it was completed, the precious cargo that made the trip all the way from Mexico City, along with the bodies of the dead teamsters and the aide, was placed in the trench and covered over. The wagons were driven away from the trail and out of sight and set on fire. The mules were all killed.

The Missourians then left the area, intending to travel to San Antonio to recruit several other ex-Confederates they knew to be living there. With a heavy guard of armed men, they would return for the treasure.

Several days later the six men arrived at Fort Concho where one of their members fell ill. He remained at the fort, promising to catch up to them in a few days when he recovered. Several days later, fully recovered, the lone ex-Confederate rode into a campsite where he found all five of his companions slaughtered, presumably by Indians. He realized that he alone knew the location of Maximilian's great treasure. He decided to travel to Missouri, relate what had transpired to friends Frank and Jesse James, and with their help return to Castle Gap to recover the gold and silver.

On his way to Missouri, the ex-soldier rode into a camp not far from the Red River and the Texas-Oklahoma border. He was invited by the three men sitting around the campfire to join them, and he readily accepted. The next morning, as the four men were finishing breakfast, the camp was raided by a posse. All were arrested and taken to jail at Denton, Texas, where the sheriff told them they were to be tried for horse theft. When the ex-Confederate tried to explain that he was only an accidental visitor to the camp, his pleas fell on deaf ears and he was summarily placed in a jail cell.

In jail he became ill, presumably with the same malady that afflicted him at Fort Concho. A doctor was summoned, the prisoner examined, and a report made to the sheriff that the man did not have long to live.

For several days the prisoner grew steadily worse and was finally told by the physician, a Dr. Black, to make his peace with God since he likely would not last through the night. The prisoner pulled the doctor close to him and in a weak and halting voice told him the account of stealing and burying Maximilian's treasure just off the road somewhere at Castle Gap. Though shaky with fever, the prisoner sketched a crude map showing the location of the treasure along with pertinent landmarks. Within a few hours he was dead.

Several days following the death of the ex-Confederate, Dr. Black told the amazing deathbed story to a Denton lawyer named O'Connor. O'Connor was interested but not persuaded until the physician showed him the map. For the next three weeks, the two men planned an expedition to Castle Gap in western Upton County in hopes of locating the amazing treasure described by the dying stranger.

On arriving at Castle Gap, Black and O'Connor discovered that the site differed markedly from the map and descriptions provided by the ex-Confederate. Though they searched for several days, they were unable to find any of the landmarks or signs as noted. Five days after arriving, the two men chanced upon the remains of several burned wagons along with the skeletons of what were apparently mules. Black and O'Connor finally returned to Denton—empty-handed.

According to one of the many versions of this tale that exist, Black passed away a few years later, and O'Connor, now a judge, returned to Castle Gap to make another attempt at locating Maximilian's treasure. He found the burned wagons and mule skeletons again but little else.

Interestingly, subsequent searches for the treasure in Castle Gap uncovered the remains of more burned wagons at no less than three different locations, each of them miles apart.

In 1929 a stranger arrived at Castle Gap with a map. When asked, he explained that his father was one of the Missouri Confederates that was involved in the theft of the treasure and the killing of the teamsters, and that the map was drawn by him just before he died.

According to the map, the treasure was buried in a long, deep trench located not far from a spring from which the Missourians drank water while they worked. Landmarks on the map directed the newcomer to a location at the south side of the gap. There was no spring in evidence anywhere, but he did find the remains of an old stagecoach station. Such locations were generally near a spring from which water could be drawn for men and horses. The spring, according to residents in the area, apparently dried up.

The stranger searched in the area for several days, found a few pieces of a burned wagon, and then disappeared. Those who came to look around after he was gone could find no evidence of an excavation and presumed he never found the treasure.

In 1968 two businessmen from Amarillo arrived at Castle Gap armed with a map and information they claimed would lead them directly to the wealth of Maximilian. After gaining permission from a landowner, they used a backhoe and excavated a long, wide, deep trench from which they intended to remove the several dozen barrels of gold and silver. Though they worked nonstop for an entire week, they found nothing.

Sometime during the early 1980s, a man was seen walking around the area of King Mountain several days in a row. He would park his car alongside the nearest county road, hike to the mountain, climb around it until dark, and finally return to his car and drive back to wherever he came from. Late one afternoon he was observed hiking back to his car carrying a heavy burlap sack across one shoulder. When asked what was in the sack, he politely displayed a number of very old silver bars he said he found in a cave on the mountain.

Hundreds of searchers have come to Castle Gap in hope of finding Maximilian's treasure. There have been dowsers, mining engineers, geologists, hobbyists, and teams of treasure hunters carrying metal detectors. Once, three priests arrived from Mexico carrying a hand-drawn map purportedly showing the location of the treasure. None of the visitors ever located any of Maximilian's stolen wealth as far as anyone knows.

Gold and silver, however, have been found in Castle Gap. Since the early 1900s, prospectors and visitors alike have occasionally turned up an ingot of silver, a bag of gold coins, or some other kind of valuable artifact, all of which fortifies the belief that more of it lies out there somewhere, just waiting for the searcher with the best map, the best technique, or the best luck.

The Treasure Coast

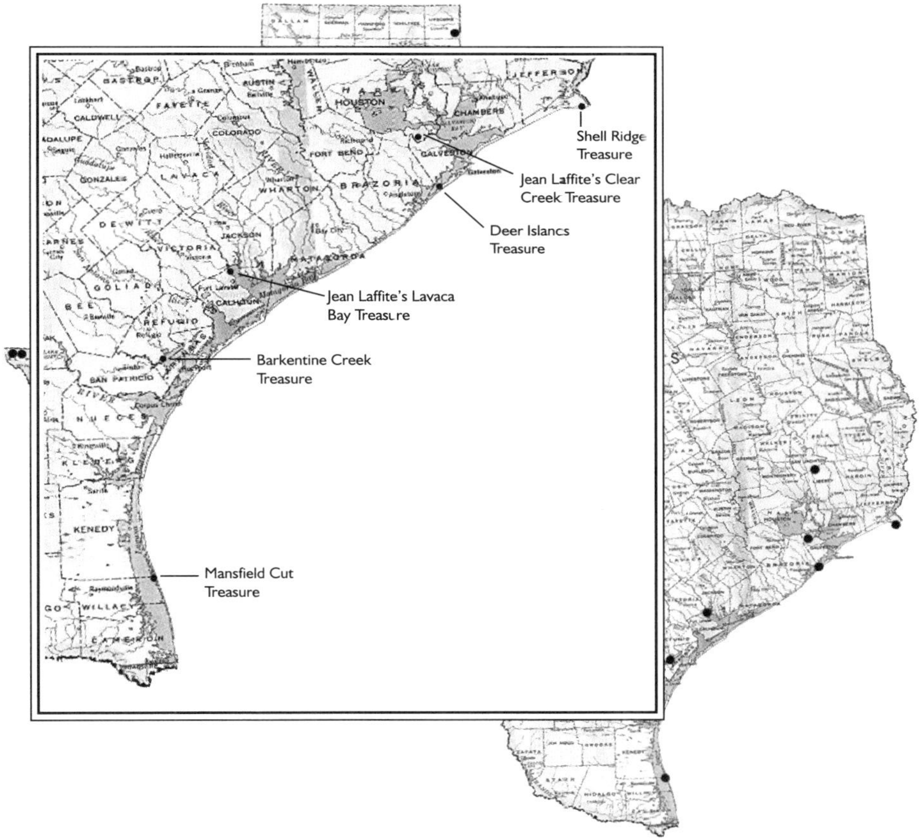

Common images associated with the Gulf Coast region of Texas include the long, clean sandy beaches that appeal to so many and fishing boats hauling in catches far out on the horizon. Many Texas residents, as well as others, visit the state's Gulf Coast each year for such activities as birding, boating, sport fishing, camping, and simply relaxing. Others, when they think of this region, think of violent hurricanes, many of which have ravaged this region throughout history.

Speaking of history, the Texas Gulf Coast is rich with it. Here, Indian tribes such as the Karankawa, the Tonkawa, and even occasional raiding Comanches could be found. These very same Indians were often an obstacle to early settlers who arrived in this area to establish productive ranches and farms. As Anglo settlement increased in the region, cities and towns sprang up, many of which eventually evolved into bustling metropolises such as Houston, Galveston, Corpus Christi, and Port Arthur.

When people think about pirates and pirate treasure, most recall tales of Blackbeard and others who roamed and ranged along the Atlantic Coast of the United States, raiding coastal towns and pillaging ships. Tales of lost and buried pirate treasure along the east coast of the United States are legion.

Less well known but equally or more fascinating are the tales and legends of pirate activity along the Texas Gulf Coast. Here, notorious pirates such as Jean Laffite lived and raided and in the process accumulated shiploads of purloined gold, silver, jewels, and other booty. Much of it was buried or lost in coastal locations, places that are searched to this day by hopeful treasure hunters. Occasionally some of this great treasure, gold and silver buried by Laffite as well as others, is found. But most of it remains lost, tempting and luring us to these coastal environs in search of the wealth and the stories.

The Mansfield Cut Treasure

Directly east of the Texas coastal town of Port Mansfield can be found Mansfield Cut—a natural channel through the 150-mile-long Padre Island. Following the many storms that strike this coast in the course of a year, a number of hopeful people rush to an area located a few hundred yards north of the Mansfield Cut to search for various treasures that can sometimes be found on the sandy beach, treasures washed ashore by the surging storm waves. Most commonly found are gold Spanish coins, but in the past items such as small silver ingots and what appear to be church artifacts have also been picked up. These objects, according to researchers, are only part of a much larger treasure, one that was being transported by a large fleet of Spanish ships four-and-one-half centuries earlier. This fleet encountered a storm much greater than the ones that occasionally wash the rich debris onto the shore, one that pummeled and eventually sank the vessels out in the gulf, perhaps three to four miles off the Padre Island shore. The bulk of the treasure transported by these ships still lies in the rotted holds of the sunken vessels, ships that still lay scattered along the gulf's ever-shifting bottom.

In 1553 a fleet of Spanish ships was being readied in the Vera Cruz harbor. The harbor, located in the thriving east Mexican coastal city of Vera Cruz, grew to become an important one for the Spanish: Virtually all ships from Spain arrived in Mexico at this location; virtually all Spanish mining and other expeditions into the interior started from this point; virtually all departures for the homeland across the Atlantic Ocean left from this point.

On a day in 1553, the vessels—twenty-three in all—were being loaded with provisions for the long journey back to Spain. As workers filled the galleys with food, made repairs to sails, and painted the decking and hulls, crews of strong workers lowered chest after chest of gold coins into the holds of the ships. This done, they added thousands of gold and silver ingots. Most of the coins and ingots were all fashioned from the products of the hundreds of rich mines that were operated throughout the interior mountains of Mexico by the Spaniards. The rest were made from gold and silver ornaments and bullion taken from the Indians. Even in sixteenth-century values, the estimated worth of the cargo transported by these ships is many millions of dollars.

The ships' captains, all experienced navigators of the oftentimes unpredictable water of the Atlantic Ocean, were in agreement that the trip back to Spain would take close to three-and-a-half months, assuming no difficulties were encountered. As was their custom, they planned to stop in Havana to replenish supplies and rest before undertaking the crossing of the vast sea.

The voyage to Havana was uneventful. After leaving Vera Cruz and crossing the Bay of Campeche, the Spaniards encountered only calm waters and light breezes. Following a week in the Cuban city, the ships once again were readied and continued eastward. Unknown to them at the time, however, a fierce hurricane was growing and heading directly into their path.

On reaching the ironically named Hurricane Flats just east of the Bahamas, the great storm struck. Several of the ships were capsized with the first impact of the storm, and many more were buffeted about so severely that masts snapped and hulls broke in two, sending the treasure-laden vessels to the bottom of the ocean. A few more of the ships somehow managed to remain afloat but were caught up in the momentum of the hurricane and blown back to the west across the gulf. In what must have been the most intensely terrifying day and night in the lives of officers and crew, an undetermined number of vessels were carried westward with the storm. Eventually the great hurricane approached the south coast of Texas, and here the rising bottom of the gulf floor caused great waves to break, violent and powerful waves which toppled over the ships that had somehow managed to remain afloat for so long. Ships, men, and cargo sunk into the shallow waters of this part of the gulf several hundred yards short of reaching Padre Island. Of the ships' occupants, approximately 300 managed to reach the shores of the island. For days the survivors foraged about the coastal flats and marshes in search of food. They lived on snakes, frogs, and an occasional heron they managed to bring down with an accurately thrown rock. Freshwater springs were found short distances inland.

Sufficiently recovered from their ordeal, the survivors fashioned crude rafts and crossed the lagoon that separated the island from the mainland and slowly, laboriously began the long walk back to Vera Cruz. Only one man, a priest, survived the tortuous 600-mile journey. Those who did not starve to death or succumb to thirst or disease were slaughtered by Indians.

Of the twenty-three ships that originally composed the Spanish fleet, three managed to break free of the great hurricane, continue their voyage across the Atlantic, and eventually reach Spain. Another limped back to the port of Vera Cruz where the captain reported what had transpired.

Nine months later, having been apprised of the report of the surviving priest, the king of Spain organized a recovery expedition to travel to the South Texas coast and retrieve as much of the sunken treasure as possible. Only a single recovery ship was sent, and it was commanded by one Captain Angel de Villafano. Within a few days of arriving at the estimated location, a number of the sunken ships were spotted approximately three to four miles east of Padre Island in the clear, shallow waters of the gulf.

The recovery of some of the treasure was undertaken, and, once the hold of Villafano's vessel held as many ingots and coins as it was capable, the captain gave the order to return to Spain. On arriving at the Spanish port months later, the treasure was unloaded and a report made to the king. A few days later Villafano was sent on another assignment, one completely unrelated to the treasure recovery. Other than his brief written report, Villafano left no maps showing the locations of the sunken vessels and no detailed directions. Years passed, and the entire matter of the sunken treasure gradually faded from the minds of everyone.

Two years before the Civil War raged throughout the South, white men were visiting the Texas coastal area and Padre Island. One of the curiosities they encountered was the fact that many of the Tonkawa and Karankawa Indians were wearing earrings and other ornaments apparently fashioned from Spanish gold coins. When asked where they got the coins, the Indians often referred to a specific location on the eastern side of southern Padre Island where, they claimed, they picked up gold coins and artifacts in great numbers from the sand.

Soon more and more white settlers were entering the region, and many of them spent considerable time exploring Padre Island. Some considered that the island might afford a good location for shipping ports, and others examined the possibilities of farming and ranching. Still others considered the opportunities for operating fleets of fishing vessels. Few of these plans ever materialized, and most of the early arrivals departed and searched elsewhere for their fortunes. A few others, however, found fortunes in gold coins and gold and silver ingots on a remote Padre Island beach.

A man named Benjamin Reno was once camped on Padre Island a few hundred yards north of a bisecting channel now known locally as the Mansfield Cut. Officially, this cut is called the Port Mansfield Channel, and it provides access to the coastal city of Port Mansfield approximately eight miles away on the Texas mainland.

One afternoon as Reno was exploring the beach, he picked up a circular object that was coated with a hard crust. Using his knife, he chipped away the crust, washed the object in the water, and was surprised and delighted to discover he held a Spanish gold coin between his fingers. A casual search of the same area yielded eleven more coins. The next morning Reno returned to the same spot and spent the entire day sifting and digging into the sand. By sundown he had collected over two hundred gold coins, twenty-one nine-inch-long gold bars, and eight seven-inch-long silver bars. In addition, he found several buttons and other trappings from military uniforms, a golden cup, and a silver cross about five inches long. In later years Reno wrote that he assumed, and correctly so, that this incredible treasure had been washed up from one or more of the wrecked and sunken ships some distance out in the gulf.

Throughout the ensuing years, dozens if not hundreds of others have discovered similar riches in the sands of Padre Island. Realizing that only a tiny percentage of the fortune had likely been washed to shore, a professional treasure hunting and salvage operation called Platora Ltd. began searching and dredging the deeper waters off the Padre Island coast in 1967. The state of Texas eventually forced Platora Ltd. to discontinue its activities, but during the peak of its recovery efforts, the company brought to the surface an undetermined number of gold and silver ingots and coins, a golden crucifix, a number of odd silver discs weighing twenty pounds or more, a cannon, several swivel guns, crossbows, and some navigation instruments.

In agreement with most estimates, the majority of researchers are convinced that ninety percent or more of the original treasure can still be found out in the waters off the Padre Island shore. If it could be located, and if one could negotiate the oftentimes confusing Texas salvage and recovery laws, one could easily be several billion dollars richer.

The Barkentine Creek Treasure Mystery

The coasts of New England and Florida have long been associated with tales of some of the world's most fascinating pirates and their lost treasures. Professional treasure hunters know, however, that per square mile, the region of the Texas Gulf Coast is likely the setting for more sunken treasure ships and more lost and buried coastal treasures than anyplace else in the world.

One particularly compelling tale involves a treasure-laden vessel that was actually swept far inland during a violent storm and ultimately mired in the soft mud of a shallow creek. The vessel, as well as the treasure it was carrying, has been seen, but today their location remains one of Texas's most fascinating mysteries.

September 1822 was remembered for many years among Spanish seaman as the "time of the storms." Indeed, during that autumn month an uncommonly large number of extremely violent hurricanes developed and grew to huge proportions far away in the Atlantic Ocean, ripped and tore westward while crossing the Gulf of Mexico, and ultimately smacked violently onto the Texas Gulf Coast in a region between Galveston and Corpus Christi. A number of vessels were lost during these storms, and dozens of others were confined to their home ports until the hurricanes were spent.

During the first week of September, a Spanish barkentine sailed out of the harbor at the Mexican port of Vera Cruz and followed a course northward across the Gulf of Mexico toward Galveston. At Galveston, the three-masted vessel was to join several others waiting there and then embark on the long voyage across the Atlantic Ocean to Spain. The cargo transported by each of these vessels was precious: wooden casks of gold coins and thousands of gold and silver ingots. All of this wealth was to be delivered to the treasury in Spain and would be used to support Spanish efforts in their quest to conquer and defend new territories around the world. The barkentine that departed the Vera Cruz harbor rode low on the sea, for every available space on the ship was filled with gold and silver coins and bars.

As the barkentine sailed parallel to Padre Island on its way to the Galveston port, the waters of the gulf began to swell and chop and the skies grew ever darker toward the east. The captain of the bark, Diego Sotomayor, kept close watch on the changing conditions, and an hour later he became convinced and then concerned that a major storm was approaching, and fast.

By the time the gulf's waves grew threatening and the winds were beginning to tear the sails from the mast, Sotomayor directed the barkentine through the narrow channel between Mustang Island and San Jose Island near present-day Corpus Christi and proceeded along the relatively calmer waters of the lagoon. Believing he had outwitted the storm, Sotomayor began to relax a bit as he contemplated the layover in Galveston.

It was not to be. The low San Jose Island and the shallow waters of the channel between it and the coastline turned out to be no protection from the violent storm whatsoever. The powerful hurricane surged across the island, blowing northwestward, and nearly capsized the barkentine. Raging onto the land, the storm dropped sheets of rain, causing rivers and creeks to swell with runoff.

On regaining control of the vessel during a brief moment of calmer wind, Sotomayor steered it into what is now Copano Bay located in Aransas and Refugio Counties. In searching for more protected waters, Sotomayor guided the barkentine into the Mission River that empties into the bay. Aided by the strong winds, the bark floated swiftly up the rain-swollen river several miles inland and turned into a likewise flooded tributary before finally becoming lodged in the muddy bottom of the shallower waters of the stream.

Two days later when the hurricane was finally spent, Sotomayor determined that over half of his crew had been lost in the storm, probably hurled overboard by the strong winds and lashing waves. He also discovered that the river's waters had finally receded and the barkentine was hopelessly mired. The only hope for returning to the gulf would be for another flood to raise the vessel from its position and carry it seaward.

Realizing any hope of reaching the sea once again in the barkentine had vanished, Sotomayor and the surviving crew members simply walked away from it, leaving the irretrievably stranded ship, and undertook the trek southward toward Mexico in hope of reaching a settlement. Mere hours from the vessel, Sotomayor and his party were attacked and slaughtered by Karankawa Indians.

Several miles back at the tributary, the tattered sails of the stranded ship fluttered in the waning coastal breeze, and the creaking of the hull could be heard from a long distance away as the vessel settled deeper and deeper into the mud. Deep inside the hold and lashed to the decks remained the incredible fortune in gold and silver coins and ingots.

Months later when the Spanish authorities learned of the fate of their treasure fleet, they assumed that all ships went down in the storm, thus a recovery expedition was never forthcoming. They knew nothing of the ship that had been lost miles inland on a tributary of the Mission River.

During the 1840s warring Comanches ranged throughout much of Texas. Not only did the Comanches seize every opportunity to attack and slay white settlers they perceived were intruding into their territory, they also attacked and looted the villages of other tribes, often taking prisoners to be used as slaves.

Around this time a party of Comanches was riding toward a Karankawa village with the intention of raiding it when they chanced upon a peculiar sight: Lying partially buried along a stream bottom they found the remains of a sailing ship. For just over twenty years the vessel had lain thus, and the rainy and humid coastal environment had taken its toll: Most of the wooden hull and deck was severely rotted and falling apart in places.

Cautiously, the Indians approached the three-masted ship. After realizing it would cause them no harm, they entered it and explored the mud- and debris-filled interior. Little in the vessel aroused much interest among the Comanches until they found some of the casks of gold coins. Though money and wealth in the form of currency and coin were not particularly important to the Indians, the Comanches knew that the white men would accept such things in exchange for rifles and ammunition.

After removing a half-dozen of the coin-filled casks and loading them onto stout horses, the Comanches were surprised by a sudden attack from a band of Karankawas. For approximately twenty minutes the Comanches rode in a southwesterly direction across the coastal prairie, fleeing before the angry Karankawas. Finally, slowed down by the heavy casks as well as an abundance of loot taken in earlier raids, the Comanches rode to the top of a low grassy knoll to make a stand. This knoll, it is believed by researchers, lies somewhat north of the town of Refugio.

After dismounting and killing several of their horses, the Comanches took shelter behind the carcasses of the animals as they fired arrows into the circling attackers. During the melee, a few of the Comanches excavated a shallow hole in the soft earth at the top of the knoll. Into this hole they placed the coin-filled casks and covered them over, intending to return and retrieve them at the first opportunity.

For more than an hour the fight raged on. During the battle, most of the Comanches' horses either ran away or were taken by the Karankawas. Having killed several of the Comanches and satisfied with the number of new horses in their possession, the Karankawas, after hurling insults at their hated enemy, rode away. When the last of them disappeared over the horizon, the Comanches gathered up what they could carry and departed. As far as is known, the participants in this skirmish never returned to the knoll to recover the buried casks of coins. At least not right away.

Approximately ten years elapsed before anyone returned to the knoll. Then in 1855 a Comanche, now old and crippled, traveled with his wife and four children to the region to try to find and retrieve the gold coins. He spent several weeks in the area but could never identify the correct grassy knoll out of all those he encountered.

Having no luck in locating the buried cache of coins, the Comanche and his family rode back toward the Gulf Coast in an attempt to relocate the old barkentine. He found it with little difficulty but noted the rotting vessel had sunk deeper into the mud. The interior of the ship was now almost completely filled with sand, mud, and debris. Reliving the time he and his fellow warriors first encountered the ship, the old Indian wandered around and across the barkentine, completely unaware that only a few feet below the dirt that filled the hold lay millions of dollars' worth of gold and silver ingots.

During the mid-1870s white settlers began moving into the Texas Gulf Coast area and establishing farms and ranches. Before long a settler named Fagan had developed one of the largest and most productive cattle ranches in the area centered in Refugio County.

Located far from forests, Fagan and others who settled in this area often found themselves desperately in need of lumber for construction. When neighbors and travelers kept referring to the existence of an old abandoned ship on a tributary of the Mission River, Fagan decided he would go investigate and determine if any of the lumber on the ship was worth retrieving.

After several days of searching, Fagan finally found what was left of the old Spanish barkentine. Leading several wagonloads of farmhands back out to the ship, he began to have it systematically disassembled, and the timbers were transported back to his farm. As he worked on the ship, Fagan had no idea that buried in the mud just under his feet was a fortune in gold and silver ingots.

Having salvaged most of the accessible wood, Fagan left the rest to decay on the coastal prairie. By 1910 the hot and humid coastal climate caused what remained to simply rot away, leaving little more than a low mound of debris. The casual rider through this area could hardly discern that a great Spanish sailing vessel once rested here.

The small tributary where the ship became mired in the mud nearly a century earlier became known as Barkentine Creek, the name by which it is known today. Somewhere in or adjacent to this creek, covered only by a thin layer of coastal prairie soil, lies one of the greatest lost treasures in the history of the Americas.

Jean Laffite's Clear Creek Treasure

Arguably, Jean Laffite may well have been the most famous pirate in history. Residents of the east coast of the United States might rather cast their votes for Blackbeard or others, but without a doubt, Laffite was certainly the most notorious brigand to ply the waters of the Gulf of Mexico. And like Blackbeard, Laffite is commonly associated with tales of great buried treasures, and there exists some compelling evidence that at least one site on the Texas Gulf Coast still contains one of this pirate's greatest caches.

For many months the pirate Jean Laffite maintained his base of operations around New Orleans and the Louisiana coast. From here, Laffite and his fleet of marauders would range out across the gulf and prey on trade and passenger vessels, accumulating millions of dollars' worth of coin, jewels, and gold and silver ingots. So great a nuisance to shipping was Laffite that the United States Navy was finally called in to patrol these waters to try to protect merchants and travelers alike.

The continued pressure applied by the U.S. Navy forced Laffite to abandon Louisiana. He sailed westward, paralleling the coast until arriving in the vicinity of Galveston Island. On landing at the island, Laffite discovered it was already being used as headquarters for a rival band of pirates. Possessing superior manpower and arms, however, Laffite simply drove the inhabitants off the island and claimed it for himself. He and his followers eventually moved into an abandoned Spanish fortress, the remains of which can still be seen just west of the Galveston Yacht Basin. From here, Laffite ranged out across the gulf, his primary targets being the Spanish treasure ships carrying gold and silver from Mexico to the Spanish homeland on the other side of the Atlantic.

Though Laffite's arrogance outraged them, U.S. Navy officials ceased to pursue him while he was attacking

Spanish vessels. Let the Spanish take care of their own, they contended. When the famous pirate returned to attacking and sacking U.S. ships, however, the navy once again went in pursuit of the notorious sea bandit, this time sending the USS *Interprise* into the gulf to pursue the pirate.

The *Interprise*, captained by an officer named Kearney, sailed directly to Galveston Island. Under a flag of truce, Kearney entered into negotiations with Laffite, hoping to convince the pirate to cease his activities and depart the area. Laffite acknowledged that two of his captains had indeed attacked American ships against his own orders. Laffite sent for the two captains and had them hung in front of Kearney, hoping to appease the naval commander. Kearney, however, was not swayed by this tactic. On returning to the *Interprise*, he ordered the ship's cannons aimed at the fortress. He then sent a message to Laffite that stated the pirate had thirty days to vacate the island. If he was not gone by then, they would blow the fortress to bits.

About one week before the end of that time period, Laffite and his men were seen loading chests, casks, and leather sacks, all presumably filled with treasure, onto his ship, the *Pride*. A short time later the *Pride* was seen sailing northward into Galveston Bay and then westward into the estuary of Clear Creek. When the waters became too shallow to allow for the draft of the *Pride*, Laffite had the treasure transferred to rowboats, which transported the booty some distance up the creek where the pirate buried it in nearly two dozen different locations along the banks.

After returning to the *Pride*, Laffite was dismayed to find his vessel had somehow become mired in the soft sands of the bay. All attempts to dislodge the ship were in vain, and the pirate grudgingly abandoned it. After loading the remainder of his goods onto his other ships, it is told that Laffite burned down his fortress and sailed away from Galveston Island. In his haste to flee from the area, Laffite left several of his sailors on Galveston Island. It is not known whether they were abandoned intentionally or unintentionally, but many of them drifted onto the mainland and undertook legitimate occupations such as fishing, farming, and ranching.

The historical record suggests the pirate Laffite never returned to Clear Creek to recover what must have been untold millions of dollars' worth of treasure in the form of gold and silver coins and ingots, as well as several chests filled with jewels.

Decades later a grizzled old seaman frequented the taverns and bars of Galveston. He was known only as Crazy Ben, and he was much like most of the derelicts that hung around seaports. The one thing that distinguished Crazy Ben from the rest, however, was the fact that he paid for his drinks and meals with Spanish gold coins.

One evening while he was in his cups, Crazy Ben told his story to a willing listener, and an amazing story it was.

As a youth, Crazy Ben served as a cabin boy to the pirate Jean Laffite himself. While Crazy Ben was not a member of the party that rowed up Clear Creek and buried the great treasure, he was aware of it, watching from the deck as the seamen loaded the casks and chests of coins and ingots into the rowboats and rowed away. He was also, he said, one of the many who were abandoned by Laffite when the pirate hurriedly sailed away. Since then he had earned a meager living on area farms, swabbing out saloons, and in general just drifting and bumming. During all this time, he said, he searched for Jean Laffite's buried treasure.

And one day he found it. At least part of it. While digging onto the north bank of Clear Creek close to the present-day community of Nassau Bay, Crazy Ben's shovel struck one of the casks. After laboriously raising the heavy object out of the hole, he smashed open the lid to find thousands of gold Spanish coins, all like the ones he now used to pay for his drinks. After pocketing several of the coins, he reburied the cask, returning to it only when he needed more gold.

For years men tried to follow Crazy Ben after watching him spend the gold so freely in the taverns, but he always managed to elude them. One night in a Galveston Island saloon, however, Crazy Ben spoke too loud and too long about his buried cask of coins, arousing the interest of several out-of-work seamen. Late that evening when Crazy Ben left the saloon, three men followed him through the dark night as he staggered toward his hovel near the bay.

It is presumed that the men who followed Crazy Ben were after not only his money but the knowledge of the location of the buried cask of coins. Whether they learned as much is not known, but the next morning Crazy Ben's body was found floating in the bay, his throat slashed.

After Crazy Ben's murder, none of the coins ever appeared in the area again, and the assumption is that his killers never learned the secret location of the buried cask. Though never substantiated, rumors have surfaced from time to time—most recently in 1999—of a few individual gold Spanish coins being found along the banks of Clear Creek near Nassau Bay.

As far as anyone knows, the remaining casks and chests containing more coins and gold and silver ingots remain buried. Experts can only guess at the value of this great treasure, but most agree that today it would be worth well over ten million dollars!

Jean Laffite's Lavaca Bay Treasure

Clear Creek is not the only location where Jean Laffite was known to have buried treasure. Another was near Lavaca Bay. Somewhere in the flats adjacent to this bay, Laffite and two of his henchmen buried a large wooden chest filled with gold coins and jewelry along with two canvas sacks packed with heavy silver bars, all booty taken from previous raids on Spanish vessels.

After Laffite was driven from Galveston Island, he led his fleet of ships southwestward, paralleling the Texas coast. After approximately 110 miles of sailing and on reaching Lavaca Bay, Laffite ordered his vessels to drop anchor while they remained here for a time before traveling on to Mexico.

During the time the pirate fleet remained at anchor here, Laffite, accompanied by two of his strongest men, loaded a large wooden chest and two large canvas bags from the ship and into a rowboat. Reputedly the chest contained a great accumulation of gold coins and Laffite's collection of his favorite jewels. Each of the canvas sacks contained a number of silver ingots.

After landing on the marshy shore, Laffite's two sailors struggled to drag and carry the chest and sacks some fifty yards inland. After selecting a location, Laffite ordered his men to excavate a deep hole and bury the treasure within. Writing in a journal, Laffite recorded adjacent landmarks, taking particular note of two separate copses of trees a short distance away. This done, he took a long brass rod he carried from the ship and plunged it into the ground immediately next to the treasure, leaving only about eight inches exposed above the surface. One version of this tale has Laffite killing the two sailors on the spot, thus leaving the pirate captain himself as the only one with knowledge of the location of the buried treasure. The most commonly related version, and the one that carries the best evidence, is that Laffite told the two seamen that he was going into hiding, and that if he did not contact them within three years to recover the treasure, then they could have it. A few days later Laffite and his fleet raised their anchors and continued sailing toward Mexico.

According to most researchers, Laffite sailed away, never to return. Where he went and what became of him has long been a matter of conjecture, but there is no evidence that he ever returned to the Texas Gulf Coast to recover any of his buried treasures.

The two sailors who helped Laffite bury the treasure near the Lavaca River went to live in New Orleans. One of them took ill with malaria and passed away within a matter of a few months. Before he died, however, he told a close friend about the buried treasure, even providing him with directions to the cache. The friend searched for the treasure from time to time for months but could never locate it. Rather than having two copses of trees to use as directional guides, he found several in the area and was unable to determine which were the ones referred to by the dead pirate.

The second sailor married shortly after returning to New Orleans, fathered two sons, and was so delighted with his family and home that he never returned to the Lavaca River. When his sons were grown, he told them the story about burying the great treasure in the marshlands adjacent to the Lavaca River. Fascinated with this tale and enthusiastic about the possibility of recovering so great a treasure, the two sons, using barely remembered directions as once related by the father, traveled to the site. They searched and searched, but like the previous hunter of the treasure, they found nothing.

Time passed and great ranches and farms sprang up along the margins of the Lavaca River, the rich grasses providing excellent graze for cattle and horses. During the mid-1800s, one of the largest ranches was owned by a man believed to be named Hill. Hill employed a black man named Owen Jones to tend his livestock as they grazed along the floodplain of the river and return them to higher ground in the evening.

Occasionally in the middle of the day and when he was certain the stock was faring well, Jones would stake his horse and take a nap. One afternoon as he was looking for a suitable spot for a nap, Jones noticed the tip of what he believed to be a brass rod sticking about eight inches above the ground. After dismounting and tying his horse to the rod, he fell asleep several yards away.

Straining against the rod, the horse had nearly pulled it completely from the soft, yielding ground when Jones awoke. Before mounting up, Jones withdrew the curious rod from the ground and carried it with him the rest of the day.

That evening when he returned to ranch headquarters, Jones showed the brass rod to the rancher. It turned out that the rancher was quite familiar with the story of the pirate Jean Laffite's buried Lavaca River treasure cache and immediately recognized the rod for what it represented. Excitedly, Hill, hopeful of finding Laffite's buried treasure, made plans to return with Jones to the floodplain the next morning. The two men arrived in the area around sunrise, and though they searched all day long, Jones was never able to relocate the exact spot from which the rod was pulled from the ground. Though he searched for the treasure in this area for the next ten years, Hill never found it.

According to material found in the research archives of professional treasure hunter and author W.C. Jameson, a number of silver ingots were accidentally discovered on the floodplain not far from the Lavaca River. They were found by a turkey hunter who was tracking a flock of the wild birds in the marshy area. Dropping to his hands and knees, the turkey hunter was quietly crawling through the grass toward a flock of turkeys when his knee struck something hard on the ground. On investigating, the hunter discovered what he initially thought to be a pile of small bricks. The hunter thought it was an odd place to find such things but he was more concerned with bagging a turkey. Before he moved on, however, he placed one of the "bricks" into his shot pouch.

When the hunter returned home that evening, he showed the curious object to his brother. After hefting it for a moment, his brother made some scratches in the surface with his pocketknife, examined them closely, and was surprised to find that the "brick" was composed of nearly pure silver. A short time later the brothers learned the story of Jean Laffite's Lavaca River treasure cache and realized the silver must have been a part of it. As soon as it was possible, they returned to the approximate site of the discovery and searched for more of the silver. For days the two men, crawling on their hands and knees, covered miles of the grassy marsh near the river but could never find the silver ingots.

During the 1920s an Oklahoma man claimed to be in possession of a journal that had once belonged to the pirate Jean Laffite. In Laffite's own handwriting, the journal contained a lengthy description of the treasure buried near the Lavaca River along with what appeared to be precise directions. The man traveled to the Lavaca River near where it enters Lavaca Bay and correctly identified, he claimed, the two copses of trees initially identified by Laffite. Unfortunately, he, like the others, was unable to find any treasure.

The marshy floodplain along each side of the Lavaca River is vast, covering several square miles and offering an incredibly wide area to search. Furthermore, the river itself, responding to the vagaries of nature, has altered its course throughout this region several times during the past two centuries. All things considered, it is more than likely that Laffite's original directions relative to the treasure cache are quite useless today.

Professional treasure hunters claim that the key to finding such a cache of treasure would likely lie in highly technological electronic equipment, perhaps employed from low-flying aircraft.

Someday, Laffite's treasure may be found, but until then it continues to lie somewhere on or in the grassy marsh near the Lavaca River, ever tempting those of us who have longed to find such wealth.

The Deer Islands Treasure

The Deer Islands consist of at least two rather unspectacular sand isles located in West Bay between Galveston Island and the Texas mainland. No one lives on these islands, and they serve as little more than stopping places for occasional fishermen. What is unique about the Deer Islands, however, is that a short distance off the coastline of one of them is the location of three sunken pirate ships, each of which, according to legend, contained huge shipments of treasure in the form of gold and silver ingots and chests of gold coins. The presence of the sunken ships near these islands has been verified several times. Getting to the treasure, on the other hand, has proven to be very difficult.

No one knows the exact date, but sometime during the month of September 1820, three privateers, fresh from raids on Spanish treasure ships in the Gulf of Mexico and the Caribbean Sea, arrived at Galveston Island.

During this time the island was a notorious hangout for pirates and was seldom visited by U.S. Navy vessels.

With a storm approaching from the southeast and the prospect of a hurricane looming large, the three vessels sailed around to the north side of the island, hoping to anchor in the calmer waters of West Bay. As the ships passed through the channel that separated Galveston Island from the Bolivar Peninsula, the storm increased in fury such that the captains were barely able to maintain a course. A short distance away on the island, they could just barely make out shacks, warehouses, and places of business being blown down as a result of the storm's fury.

After positioning the ships as close to the stone pier as possible, anchors were dropped, rowboats lowered, and with great difficulty, the pirates made their way toward the shore and, they hoped, some degree of protection. In their haste to abandon the storm-tossed vessels, they neglected to batten them down securely such that they might ride out the storm. Furthermore, as they were anxious to reach shore and seek shelter from the violent storm, they opted to leave the cargo of gold and silver bars and coins aboard the ships.

On landing, the pirates found a stone warehouse nearby that resisted the high winds, and toward this they made their way. Once inside the warehouse, they encountered dozens of citizens of the small community, all enjoying the protection of the stout rock walls against the storm.

Despite the hurricane that was raging outside, the pirates walked in on a loud, brawling party, for in the warehouse was stored a quantity of rum that had been discovered by the occupants. Before long everyone in the warehouse was drunk and completely unaware of what was transpiring outside in the harbor.

As the merriment continued in the warehouse, the powerful storm winds tore the ships from their moorings and pushed the vessels northwestward out into the bay. Soon the three ships were blown out to North Deer Island. As they careened across the shallow sandy bottom, they broke open on impact and one by one sank into the soft silts just a few feet below the surface.

It was a full forty-eight hours later when the storm finally passed and the sailors recovered from their prolonged revelry. It was only when they walked out onto the pier that they realized the ships were gone. Commandeering two other vessels that had remained successfully anchored in the harbor, they sailed up and down the shore searching for their own ships but could not find them. Having no idea that they could have been blown as far to the north and west as the Deer Islands, the pirates never considered looking in that area. After several weeks of fruitless searching, they finally gave up. During the ensuing months, the sailors were sent on other missions and raids, and eventually the three missing ships were forgotten.

It was easy for the three sunken treasure ships to remain forgotten. During that time hardly anyone ever visited the Deer Islands. The low-lying, bare, sandy islands held little attraction for anyone other than an occasional fisherman who might pause in the area for a time. Unknown to them, the three treasure-laden vessels lay mired in the bottom sands and silts mere inches below the surface of the water.

In the nineteenth and twentieth centuries, Galveston Island and the adjacent mainland grew more and more populated as enterprising settlers sought opportunities related to farming, ranching, and commercial fishing. West Bay became a favored spot both for commercial and sport fishermen, and the waters around the Deer Islands were being visited more and more frequently.

During the spring of 1953, a lone fisherman rowed his wooden boat out to a favorite spot on the north side of North Deer Island. Without warning, a sudden storm hit the area. Rather than try to row the comparatively long distance back to the mainland, the fisherman guided his small boat toward the island. After beaching his boat, he sought what little shelter was available from a low growing tree he found near the shore. Crouching beneath it, he had little choice but to wait out the storm before returning home.

About four hours later when the storm finally weakened, the fisherman walked out onto the mud flats to assess the condition of the lagoon. The first thing he saw gave him a start. There, in the water a little over one hundred yards from the shore lay the hulks of three sunken ships. Along the port side of one of the vessels the fisherman could discern a row of what appeared to be cannons. Intrigued at the prospect of finding long lost sunken wrecks, the fisherman began wading out across the flats toward the vessels, but he sank deep into the soft mud and was forced to return to shore.

After reaching the sandy beach, he retrieved his own boat and attempted to row out to the wrecks, but the water here was so shallow he was unable to successfully negotiate the mud flats. So excited was he about the prospect of finding something of value on the vessels that he took neither compass readings nor note of pertinent landmarks. It would soon prove to be a grave mistake.

Hours later he returned to the mainland and immediately began organizing a return trip to North Deer Island. Three days later he arrived at the location where he was certain he saw the sunken ships, but nothing was to be found. An examination of the coastline revealed little that was familiar, so he explored along the shore, searching for the tree under which he took shelter during the storm. Try as he might, he was unable to locate it. For several days he searched the area but was never able to find the three sunken ships.

Since that time, occasional fishermen and low-flying pilots have reported spotting one or more sunken ships in the area of North Deer Island. They are variously identified, but most of the reports indicate that they appear to be very old sailing ships and have likely been in the water for well over a century. A number of organized expeditions have set forth in search of them, but the mysterious vessels continue to elude the treasure hunters.

In 1970 a Houston, Texas jeweler received a curious visit from a man who showed him an eight-inch-long ingot of silver he said he would like to sell. The jeweler examined the bar, proclaimed it to be almost pure silver, and asked the man if he had more of them. The stranger replied he was in the possession of several and that he found them among the wreckage of some old sunken ships in the mud flats of North Deer Island.

After paying the stranger for the bar, the jeweler asked him when he could bring in the rest of them. Without replying, he simply stuck the money in his pocket and scurried out the front door. The jeweler never saw him again.

Several days later the jeweler related the tale of his strange encounter to a fellow businessman. Intrigued, the businessman went to what he presumed was the approximate location of the alleged find at North Deer Island only to discover that much of the area had been recently dredged as a result of the construction of the Intracoastal Waterway, a ship channel that today winds through the lagoon between the Texas mainland and the outer islands. Huge mechanical dredges were scooping up tons of mud and sand and depositing them in a series of banks just north of the channel. Investigating one of the banks, the businessman found what he believed were rotted planks from ships that had been under water for a long time. He also found a crude astrolabe, a navigation instrument commonly used over a century earlier.

Undoubtedly, he concluded, the mechanical dredge had scooped up the remains of the three sunken ships, along with their contents of treasure, and simply incorporated them into their spoils banks. According to researchers, untold millions in gold and silver coins and ingots are likely buried in and among the muck and sand.

This was not received as bad news by professional treasure hunters, for they claimed that mining the spoils banks would be much easier than trying to recover any real or potential treasure from the treacherous mud flats of Deer Island.

As this is being written, plans are currently underway by a team of treasure hunters to electronically examine a portion of the spoils banks in order to determine the location of any gold and/or silver.

Shell Ridge Treasure

Just a few dozen yards off the Texas shore in the Gulf of Mexico can often be found a number of small sand islands. These islands are of an ephemeral nature: Due to the vagaries of wind and water currents, they sometimes appear overnight and may remain for years. Others suddenly vanish, washed away by storm waves and winds. Many of the tiny sand islands are seldom around long enough to be given names, but one of them, located near the Texas-Louisiana border, is known by the locals as Shell Ridge. The reason it is recognized and known to many is that, according to legend, a great pirate treasure is buried somewhere on it, a fortune that has attracted searchers to this area since the early part of the nineteenth century.

Henri Lapierre had spent most of his life aboard a sailing ship. As a young boy he was forced into service on a pirate ship and made to clean decks and help the cook. Young Lapierre joined in the raiding and fighting and eventually proved himself to be worthy of joining the ranks of the feared brigands who plied the high seas and robbed British and Dutch trading vessels.

In time Lapierre fell in with the famous pirate Jean Laffite and became one of his chief lieutenants. Lapierre was with Laffite when the notorious brigand was removed from Galveston Island by the U.S. Navy. Rather than follow Laffite to Mexico, his intended destination, Lapierre took command of several abandoned ships and decided to command his own fleet of pirate ships.

Lapierre had numerous successes, but none as great as the capture and pillage of a treasure-laden Spanish galleon in the Caribbean Sea. Following the sacking of the vessel and placing the treasure, which consisted of gold coins, several gold ingots, and hundreds of jewels and uncut emeralds, into the hold of his command ship, Lapierre set sail for New Orleans where he planned to rest and spend some of his newfound wealth.

On the way to the Crescent City, Lapierre's fleet of five ships was caught in a hurricane. For hours the pirates, experienced and able seamen all, fought the violent winds and surging waves, but it was a losing battle. When the winds finally ceased, Lapierre discovered that all of his ships save the one he commanded had sunk. His own vessel was seriously disabled—all three of the masts had snapped near the base and he was without sail. Furthermore, his rudder was broken. When he counted his sailors, he also discovered that several had apparently been swept out to sea. Unable to guide the path of the ship, Lapierre and his crew drifted with the currents for three weeks before spotting land.

Helpless, the pirates stood elbow to elbow on the starboard side of the ship as it floated toward a small sandy island about a quarter mile from the mainland. Finally the ship caught on the shallow flats that extended gulfward from the island and began sinking into the soft, muddy bottom.

Above all else, Lapierre wished to save his men and his treasure. He ordered his crewmen to begin carrying the treasure, which had been stored in metal buckets and leather bags, from the hold and placing it into the lifeboats which had been lowered into the water. From his own quarters, Lapierre secured a large trunk that had been used to store his clothes. The trunk and the treasure were then rowed to shore and carried to the middle of the small island.

After dragging the trunk and the several bags and buckets of treasure to the approximate center of the island, Lapierre ordered his crew to excavate a deep hole in the sand. This done, the pirate leader placed the large trunk into it and then commenced filling it with the contents of the bags and buckets. After refilling the hole, Lapierre had his men place two large pieces of driftwood they found across the top. Into each of the driftwood logs, Lapierre hammered a large brass spike.

Leaving the island, the pirates, fourteen in all, rowed to the mainland. During the short trip, one of the rowboats sprang a leak and sank. The three sailors in the boat were unable to swim and drowned before their companions could reach them. On reaching the shore, two more sailors got into a fight, and before they could be separated, one was lying dead on the beach, a knife protruding from his chest.

From this point, the ten remaining pirates, including Lapierre, began walking eastward with the intention of eventually reaching New Orleans and organizing an expedition to return to the small island to recover the treasure. Before leaving, Lapierre made note of pertinent landmarks. The island, he determined, was less than a quarter mile from the mainland and just a short distance east of where the Sabine River flows into the Gulf of Mexico.

The long journey on foot to New Orleans was fraught with hazard and tragedy. Two of the pirates became trapped in quicksand and sank out of sight before anyone could reach them. Four more were killed by hostile

Indians, and the remaining pirates barely escaped with their lives. Three simply wandered off and were never seen again. Weeks later only Henri Lapierre and his first mate made it to New Orleans.

During the weeks that passed while the two pirates were recovering from their ordeal, the first mate got into a fight in a local tavern and was killed, leaving Lapierre as the only person left alive who knew the location of the buried treasure.

Lapierre was without funds and weakened from the long and hazardous journey through the malaria-infested swamps of southern Louisiana. Four years passed before Henri Lapierre was able to organize a return trip to the small, sandy island to recover the treasure. Finally, he left New Orleans alone and traveled overland, eventually arriving at a point along the Texas coast where he could see the familiar island a short distance out in the gulf.

Not far from Sabine Pass, the channel through which the river flowed to reach the gulf, Lapierre encountered what appeared to be a ranch. After walking along a lane, he eventually came to a house, knocked on the door, and inquired about securing a room for the night.

Lapierre was greeted by a Mrs. Gaffney, who assured him that, indeed, she did have an extra room and that it was available. That evening, rancher Neil Gaffney, along with his two sons, returned home from their chores and met and visited with the stranger.

The next morning after breakfast, the newcomer asked Gaffney if he had a skiff to rent. He told the rancher he wanted to row over to the little sandy island just offshore. Gaffney assured him he did have a rowboat he was welcome to use and told him the name of the island given to it by the local residents was Shell Ridge.

Each of the next three mornings, Lapierre rowed to Shell Ridge. He walked about the island all day long, stopping long enough to probe the sands with a long pole he carried. Now and again he would stop and dig a hole using a shovel he borrowed from Gaffney.

At the end of the third day, Lapierre decided to explain the purpose of his visit to the Gaffneys. As the family and the former brigand sat around the table at dinner, Lapierre told them he had been a pirate with Laffite, about taking the treasure from the Spanish galleon, and about burying it in a large trunk on the island. He also told the family about placing the two driftwood logs over the site and hammering brass spikes in each of them. He then told Gaffney if he would help him find the treasure he could have half of it.

For several weeks the two men combed the island looking for the treasure. Severe storms that had struck the region during the four years Lapierre was in New Orleans had reshaped the island somewhat. Lapierre presumed surging waves washed the logs away.

One morning Gaffney informed Lapierre he needed to travel to New Orleans on business, and he invited the ex-pirate to join him. After spending two days and nights in the Crescent City, Lapierre booked passage on a boat bound for New York City, explaining to Gaffney that he intended to meet with a man who might help fund an ongoing search for the treasure. He promised he would return in a few weeks. The two men said goodbye, and Gaffney returned to his Sabine Pass ranch.

Months passed, and Gaffney never heard from Lapierre. In fact, no one knows what became of the former pirate, and he was never heard from again. With

Lapierre gone, Gaffney soon lost interest in searching for the alleged treasure on Shell Ridge.

One morning during late autumn in 1831, Gaffney rowed his skiff out to the gulf side of Shell Ridge. He was collecting firewood, and he knew that the waves often washed driftwood up onto the shore of the tiny island. The storm had also eroded away a portion of the island and removed a great deal of sand from the middle of it. Gaffney was in luck, for a recent storm had deposited hundreds of pieces of driftwood onto the shore.

As Gaffney was gathering the driftwood, he spotted a length of log several yards away on the island. Curious, he walked over to it and found the end of a brass spike projecting from it. Recalling Lapierre's story of how he marked the location of the buried treasure with two such spikes, Gaffney began digging away the sand under the log. Minutes later he struck the top of a very large wooden chest. Exerting greater effort, he removed even more sand, finally exposing several inches of the chest.

Using a metal bar he retrieved from the boat, Gaffney pried open the lid to the box and beheld an incredible sight—lying in the chest were thousands of gold coins, hundreds of uncut emeralds, and precious jewels fashioned into rings, necklaces, and broaches and fastened into settings of fine gold and silver. Gaffney realized he was looking upon a fortune ten men could not spend in a lifetime.

After covering over the treasure-filled chest, Gaffney, his rowboat filled with driftwood, returned to his house. He spent the remainder of the day in his shed making five wooden boxes, each one approximately eighteen inches by fifteen inches, each fitted with a hinged lid that fastened tightly.

The next day he returned to Shell Ridge, uncovered Henri Lapierre's treasure chest, removed the treasure, and after dividing it up placed it into the five separate boxes. Following this, Gaffney buried the boxes in five separate locations on the island, each one several hundred yards apart. Gaffney did not place locational markers on any of the sites, believing he would be able to return to exact locations of the buried chests. Gaffney reasoned that if anyone came to the island searching for Henri Lapierre's treasure and uncovered one of the boxes they would presume they had it all and would not search for the other four.

When Gaffney returned home that evening, he told his wife about the treasure he reburied on Shell Ridge. During the next few weeks, he told her, he, along with his hired hand, would round up all of the cattle, drive them to New Orleans, and sell them. Following this, he would return to Shell Ridge, retrieve the treasure, sell the ranch, and move to Ireland, his homeland, where they would live in wealth and splendor.

It was January of 1832 by the time Gaffney's entire herd had been rounded up. This done, he and the hired man made the long drive to New Orleans where he sold the cattle. Gaffney paid off his herder and told him he would no longer be needed.

During the cattle drive to New Orleans, Gaffney came down with a high fever. While he was conducting business in the Crescent City, he was continually bothered with the fever along with a nagging cough. Though quite ill, he was determined to finish his business and return to his ranch in Texas as soon as possible.

On the return journey, Gaffney became so ill that by the time he reached the small bayou settlement of

Franklin west of New Orleans he was unable to remain on his horse. He was found lying by the roadside and taken to a local woman known to treat sicknesses with herbs and potions. Her doctoring skills were not adequate to save Gaffney, however, and the rancher died two days later.

When Mrs. Gaffney learned of her husband's death, she mourned for several weeks and then decided to try to find the five buried treasure boxes on Shell Ridge. With her children, she rowed out to the island every morning and searched all day long. She searched every day for weeks but never found anything. Mrs. Gaffney continued to live on the ranch near Sabine Pass for the remainder of her life. Even as an elderly woman she sometimes returned to Shell Ridge to try to find the treasure reburied by her husband, but she never did.

Today it is difficult to determine which of the tiny sand islands near Sabine Pass is actually Shell Ridge, if any of them. It may have long since been eroded away by wind and waves, or it may still be where it was when Henri Lapierre buried his treasure on it well over a century and a half ago.

THE BEST OF THE REST

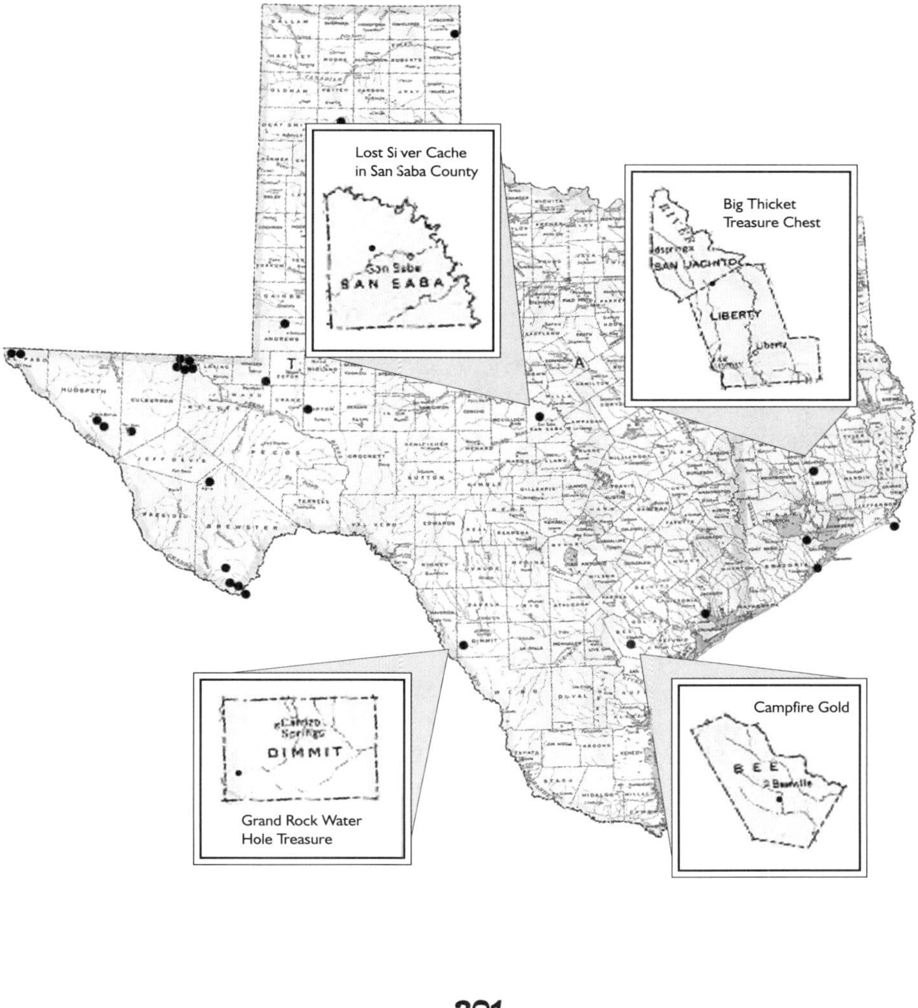

To be sure, West Texas, including the Guadalupe Mountains, the Big Bend region, the Panhandle, and much of what is called the Trans-Pecos area remains one of the richest areas in the world as far as providing settings and circumstances for tales and legends of lost mines and buried treasures is concerned. Additionally, though never receiving the amount of attention given the Atlantic Coast and Florida, the Gulf Coast of Texas has been an incredible source for fascinating stories of pirate treasure, and it remains a promising location commonly visited today by professional and amateur treasure hunters as well as writers and researchers of such aspects of our history and culture.

The rest of the Texas regions—the Cross Timbers region, the Piney Woods, the Big Thicket, the Blackland Prairie, and the southern part of the state—do not manifest the rich natural outcrops of gold and silver and other minerals as are often found farther to the west, but they remain intriguing locations for tales of lost and buried treasures. Though the treasure tales encountered here are not as numerous as those that have emanated from farther west, they are nonetheless just as compelling and fascinating, if not more so.

The following are among the best of the stories collected from North, Central, and East Texas.

Lost Silver Cache in San Saba County

For several weeks following the end of the bloody War Between the States, the citizens of the state of Texas kept busy recovering from the turmoil and trying to settle back into a pattern of peaceful farming and ranching. San Saba County, located some 150 miles southwest of Dallas was not impacted heavily by the war, but as with most other places, animosity ran high and strangers were generally regarded with suspicion.

Thus it was when Charo Sanchez rode into the town of San Saba. Sanchez, who was close to seventy years old at the time, rode an aged, swaybacked mule that looked as if it might drop dead at any moment. Having little or no money, Sanchez set up a small camp near the edge of town and rested for a few days.

One afternoon after staking out his mule to graze on the grass that grew near the camp, Sanchez entered the town and began asking questions concerning the location of an old furnace several miles to the south. The furnace, he said, was used to melt down ore before casting it into ingots. No one he encountered in the town was aware of the existence of the furnace, and Sanchez was turned away each time.

Several days after Sanchez' arrival, a rancher drove a wagon in from the west and pulled up just outside the mercantile store. The rancher had returned to his holdings after serving in the Confederate army and was anxious to reestablish his cattle herd. As he was loading supplies into the wagon, Sanchez approached him and requested permission to ask a few questions.

The rancher, seeing only a poor old man clearly down on his luck and with no money, agreed to have a conversation after his wagon was loaded. Sanchez offered to help, and soon the supplies were stacked neatly in the vehicle. The rancher was impressed with how hard the old man could work. When the job was completed, the rancher took Sanchez to the local café and bought him lunch in gratitude. As they dined, Sanchez told the rancher a most amazing tale.

Thirty years earlier, related Sanchez, he and a partner, along with twenty laborers, were digging silver from a mine located close to the Little Llano River located a few miles to the south. After removing the ore-filled rock from the mine, they would carry it to a crusher over a mile away and close to the river where the silver would then be separated from the matrix. While the workers were all occupied with crushing the rock, they concealed the opening of the mine so it could not be seen by passersby while they were absent.

From the crusher, the silver would then be taken to the furnace where it was melted and poured into eighteen-inch-long ingots. At the end of each month, the accumulated ingots were buried. After six months of labor, said Sanchez, it was estimated that well over 2,000 pounds of ingots, along with approximately $500 in Mexican gold coins, were cached in a location approximately 400 yards north of the furnace and not far from a spring.

Sanchez explained how they had just completed burying a significant amount of silver at the end of the sixth month when the party was suddenly attacked by Indians. So unexpected was the attack that the miners were caught completely unprepared and ill-equipped to fight back. As a result, all but Sanchez, his partner, and a young girl were killed. The girl, barely sixteen years of age, was the wife of one of the workers. She often cooked and cleaned for the miners. During the first few moments of the attack, the three crawled into a dense thicket of brush and lay quietly hidden while the slaughter ensued. In horror, the three watched as the miners were stripped of their clothes and boots, scalped, and then mutilated. After the Indians departed with all of the horses and mules, the three remained in hiding until sundown. Then quietly crawling out from their place of concealment, they hurriedly packed what useful items they could find scattered on the ground and fled southward to Mexico, arriving in Mexico City several months later.

Because of their ordeal and their fear of the hostile Indians, Sanchez and his partner delayed their return to the area to retrieve the huge cache of silver. They decided to allow the passage of at least two years before undertaking a return trip.

During that time, the partner and the girl were married. At the end of two years, she was pregnant with their second child, so they decided to wait until several weeks after the birth to leave for the mine. When the time arrived to depart, Sanchez was stricken with a severe fever and the trip was postponed once again. For several weeks Sanchez was abed with the sickness, and his friend feared for his life. Eventually he recovered only to be met with another setback.

As they once again made plans to return to the San Saba region, the partner was thrown from a horse and suffered a severe fracture of his right leg. As a result, many more months passed before they spoke again of returning to Texas.

Finally, carrying several weeks' worth of supplies loaded onto stout pack mules, Sanchez, his partner, and his partner's wife, undertook the long journey back to the silver mine in San Saba County. Unfortunately, they were attacked by bandits in the northern part of the state of Chihuahua. Though their lives were spared, all of their money, supplies, and animals were taken. Dejected and penniless, they returned to Mexico City.

More years passed, and the three embarked on another attempt to return to the silver cache. It was agreed that Sanchez would journey to San Saba County while the partner and his wife, traveling via stagecoach and rail, went to St. Louis where they intended to purchase a home and settle down. When Sanchez located the cache and the mine, he would send for them and together they would retrieve the silver.

Upon arriving in San Saba County, Sanchez, having been away from the region for so many years now, recognized very few landmarks and became lost time and again. Though he spent several weeks in the area, he was unable to relocate the mine or the cache. Frustrated, he decided to travel to St. Louis where he planned to locate his partner and, together, they would go to San Saba County to claim their wealth.

When he arrived in St. Louis, Sanchez looked up his partner, and as he spent several days resting from his journey, the two men made plans to revisit San Saba County. On the day before their intended departure, the partner was mounting his horse when his pistol accidentally discharged, killing him.

After burying his partner, Sanchez returned again, alone, to San Saba County to search for the cache and the mine, but luck was not with him. Finally he returned to Mexico.

Now an old man, Sanchez told the rancher that he had lived a full and relatively happy life, had outlived three wives, and had many children, all grown and gone. It was his wish, he said, to make one last attempt at finding the fortune in silver he knew lay buried not far from the Little Llano River.

When he could spare the time, the rancher aided Sanchez in his search. Though the two men looked for the cache off and on for nearly three months, it was never found. Bidding his new friend goodbye, Charo Sanchez returned to Mexico. Several weeks later the rancher learned Sanchez had passed away quietly in his sleep.

For the next several years, the rancher, convinced of the truth of Sanchez' story, searched for the cache and silver mine when time permitted but never found either.

During the late 1870s, a San Saba County sheepherder named John Medlin found the old ore furnace. Medlin, who had long been familiar with the story of the lost silver mine and ingot cache, spent much of his time looking for them while grazing his sheep near the Little Llano River. One day he chanced upon the weathered and tumbled-down ruins of the little smelter. During the next several weeks, he ranged northward from the smelter in search of the spring near which the cache was located. His efforts were finally rewarded one afternoon when he came upon it purely by accident. Though he dug several holes in the area of the spring over the next few days, he never found any silver. During his excavation efforts, Medlin did find several human bones, all presumed to be the remains of the miners who were killed during the Indian attack.

In 1884 Medlin and his family moved to California. Before leaving, he related the story of Charo Sanchez' lost silver cache to Nancy Bower, the young daughter of an area rancher. Years later after Bower was grown and married, she undertook a search for the cache. Like Medlin, she found the remains of the old furnace and the spring. Encouraged by this, she spent many weekends digging in the ground near the spring. One afternoon she was finally rewarded with the discovery of three eighteen-inch-long silver ingots she encountered only an inch or two below the surface. Excited, she continued her efforts for several months, but she finally gave up after finding little else save for some old rusted tools.

The search for this lost San Saba County silver cache continues to this day. In 1954 six more silver ingots were found near the spring, but the main cache, believed to contain well over 2,000 pounds of ingots, continues to elude the treasure hunters.

Grand Rock Water Hole Treasure

The southwestern corner of Dimmit County lies just a few short miles from the Rio Grande in South Texas. Located in this county is a well-known spring, an oval-shaped pool of clear, cool water called the Grand Rock Water Hole. For centuries Comanche Indians camped near the spring during their wanderings through this part of Texas. Early Mexican and white travelers and settlers knew about the spring, and though it is located in a comparatively remote part of the Lone Star State, it was widely used by those who visited this region.

It was long believed that the Grand Rock Water Hole was bottomless. Indeed, there are those living in the nearby towns of Carrizo Springs and Crystal City who maintain it is to this day.

It is also believed by many that an incredible fortune in gold and silver lies somewhere deep within Grand Rock Water Hole, a treasure placed there during the early part of the nineteenth century by a party of Mexicans, and one that remains deep within the hole, one searched for by professional and amateur treasure hunters alike. To this day the treasure, estimated to be worth several million dollars, has never been recovered.

During the first decades of the 1800s, Texas was a Mexican possession and San Antonio was its largest city. A number of wealthy Mexicans moved to San Antonio to take advantage of its center as a commercial hub and crossroads for the numerous pack trains that bore trading goods throughout much of the region. Business was brisk and growing, and soon hundreds of thousands of dollars' worth of goods and money were on the trail between San Antonio and Mexico City at any given time.

During the last week of June in 1803, a large pack train carrying what has been estimated at several million dollars' worth of gold and silver in coins and ingots was being readied in Mexico City for the long journey north to San Antonio. Dozens of mule drivers were employed to take the pack animals northward, and over fifty armed guards were assigned to protect the train from bandits and Indians who often frequented the road.

At the helm of this grand voyage was one Captain Palacio Flores. Flores, an academy graduate from a prominent family, was well connected with Mexican politicians and officials and was a favorite officer of General Santa Anna. Many predicted Flores would eventually find a position of importance in the growing government of Mexico. Some predicted he might even become president someday. His status among the country's decision-makers led to his appointment to this extremely important task—deliver the wealth safely to San Antonio.

For several weeks the slow-moving pack train plied its way across the pleasant hills and valleys of eastern Mexico, pausing to rest men and animals alike at locations such as Celaya, San Luis Potosi, and Monterrey. The trip was, by and large, uneventful, for not a single Indian or bandit was encountered along the way. Flores and his cadre of officers were growing confident. With their increasing confidence, however, came carelessness.

The treasure-laden pack train forded the Rio Grande at a barren and remote crossing approximately fifty miles upstream from the settlement of Laredo. From here the road wound through arid, rocky, brush-covered hills.

Within an hour after crossing the river, one of the guards reported to Flores that the pack train was being watched by some men on horseback. Scanning the low ridge about a half-mile to the east, Flores spied about thirty horsemen, all standing still, all focused on the pack train.

Flores ordered the pack train forward, and as it snaked its way along the rough, narrow road, the armed guards and muleteers nervously watched the horizons in anticipation of an attack. Hours passed and the newcomers made no move, but an hour before sundown, yet another of the guards reported to Flores that the train was being followed by the same group of riders who had watched earlier from the ridge. A short time later Flores called for a halt and ordered the gold and silver unloaded from the pack animals and stacked in the center of camp. Around the treasure he stationed guards who would watch over the treasure in shifts throughout the night. Other guards were placed at selected locations around the perimeter of the camp. It was a nervous and frightened party of Mexicans that evening, and very few of them slept at all. The night passed without incident, however, and the men rose, breakfasted, reloaded the treasure, and saw the pack train underway along the road before the sun was two hours above the horizon.

Owing to the rocky and badly eroded road, travel was especially difficult this morning and the train made poor time. San Antonio still lay 120 miles to the northeast, but at this rate Flores feared they would not arrive in two weeks. Though they made no move to approach the pack train, the mysterious riders spotted the previous day still trailed approximately two miles behind. According to a scout that observed them, their numbers had increased to approximately forty, and all appeared to be well armed.

The road followed Pena Creek for several miles, and since the weather was turning quite warm, Flores allowed the men and animals to stop frequently to rest. Late in the afternoon of the second day after crossing the Rio Grande, Flores was growing more and more concerned about the slow rate of travel and the proximity of the unidentified men behind the pack train. Consulting a map he carried, Flores directed the pack train toward a campsite a short distance away, a location known as the Grand Rock Water Hole.

On finally arriving at the spring, Flores once again had the pack mules unloaded and the great treasure piled only a few feet away from the water hole, a five-foot-wide pool of deep, clear blue water. Seeing that the graze was comparatively good here, he instructed his soldiers and muleteers to stake out the horses and mules and that they would remain encamped at this location to rest up sufficiently for the final leg of the journey into San Antonio. All through the night, Flores stalked around the perimeter of the camp, checking on the guards, which he doubled, and making certain that all was in readiness should an attack from the mysterious riders materialize. Flores finally fell asleep two hours before dawn, confident that no bandits could pass through his well-trained sentries.

The next day Flores sent scouts out to determine the location and number of men following the pack train. Around midday the scouts returned and reported they had encountered no sign of the riders at all. Flores decided that the trackers, on seeing the well-armed and well-trained guards with the pack train, simply decided not to try to commit a robbery and rode away. That night, the second night at the campsite, Flores did not even double the guards.

It was a serene and rather complacent group of men that easily fell asleep that night, each convinced that the threat of bandits was removed. Even a number of the sentries fell asleep at their posts. Little did anyone know that throughout the night forty mounted outlaws were taking up positions in the low ravine adjacent to the camp.

Just as the sun broke over the horizon the next morning and some of the muleteers and guards began to stir, the thundering sound of hoofbeats overwhelmed the sleepy soldiers and muleteers as the bandits rode out of the ravine and charged the camp. Taken completely unaware and unprepared, at least twenty guards and mule drivers were killed at the first charge.

By the time the bandits readied themselves for a second rush, Flores was positioning his soldiers in a skirmish line in an attempt to provide some resistance to the attackers. As muskets were being primed, Flores, determined that the bandits would get none of the treasure, ordered the muleteers to commence tossing the packs and crates of gold and silver into the deep waters of the spring.

There was little else the captain could do. To allow the great wealth which had been entrusted to him to fall into the hands of mere highwaymen would be a supreme embarrassment and cause his career as a military officer to be ended. Following the defeat of the bandits, which he believed would be imminent, he would simply have his men recover the gold and silver, and they would continue on their way to San Antonio.

The next charge by the bandits caused Flores to have second thoughts. The relentless onslaught by the determined highwaymen resulted in the killing or wounding of the entire front line of Mexicans. After seeing that the last of the treasure had been thrown into the pool, Flores turned toward the attackers, pulled his sword from the scabbard, and joined his countrymen in the front ranks.

A few minutes of silence reigned as the bandits regrouped some distance from the camp. Flores glanced around at his command and saw that most of them were now dead. His heart sank as he realized they would not be able to hold out much longer. Trying vainly to appear confident, Flores walked among the men trying to encourage them, telling them that victory was only moments away. The look of hopelessness in their eyes revealed they no longer had any confidence in their commanding officer.

The third charge was the final one, and Flores was one of the first to fall from an enemy bullet. Mortally wounded, he collapsed across the bodies of three slain soldiers. Within seconds the bandits were riding among the Mexicans, shooting and slashing. While most fell to the bullets, sabers, and knives of the highwaymen, about half a dozen fled, taking shelter in the nearby ravines and woods.

One of the mule drivers, a man named Alejandro Lucero, ran into a grove of trees, climbed a thick oak, and concealed himself among the leaves. As he watched, the bandits rounded up the horses and mules and rifled through the dead men's pockets. Several of the bandits gathered at the water hole and engaged in animated discussion. Presently two of them stripped off their clothes and dove into the water. Repeatedly they submerged, but came up each time with nothing. After a while the two divers climbed out of the water and dressed. An hour later the bandits, herding the purloined horses and mules ahead of them, rode away.

When he was certain it was safe, Lucero climbed down from the tree. Following a brief examination of the field of battle, he then proceeded on to San Antonio afoot.

Several days later a teamster found Lucero on the roadside only a few miles from San Antonio. After placing the starving and dehydrated man into the wagon, the teamster carried him into town and delivered him to officials. When he finally regained his strength, Lucero related the events of the massacre at the Grand Rock Water Hole. One or two highly ranked officials recommended forming a large and well-armed force to go in pursuit of the bandits, but there was little enthusiasm among their followers to do so. Others wanted to make an attempt to retrieve the treasure from the spring, but most expressed fear that the bandits were still close by and did not wish to encounter them.

Weeks passed, and from time to time word arrived in San Antonio that the gang of bandits continued to prey on travelers along to the road to Mexico. What enthusiasm remained for conducting an expedition to the Grand Rock Water Hole and making an attempt to retrieve the treasure gradually waned. Months later the topic was relegated to one of minor importance, and after several years had passed, virtually all had forgotten about it.

Ninety years after the Grand Rock Water Hole Massacre, a San Francisco newspaper, while researching some archival material, found an account of the caching of a large amount of gold and silver at a waterhole in Texas near the Mexican border. Investing a great deal of time and energy, the reporter located other documentation that revealed the names of the parties to which the treasure was to be delivered. Using this, he identified a number of descendants of the listed recipients, contacted them, and was instrumental in organizing a salvage operation in an attempt to recover the treasure.

Since it had been a long time since this portion of the San Antonio to Mexico road had been used, locating the spring proved to be very difficult. When it was finally found, it was exactly as described in the archival material: delicious, cool, and very clear water. Cutting long poles from some nearby trees, members of the party probed for the bottom but never made contact. One man, who identified himself as an excellent swimmer, dove into the pool. After several attempts to reach the bottom, he finally came ashore and claimed he could not see the bottom and that its depth was much farther than a man could dive.

Several ropes were tied together with one end weighted by a large rock. The rock was lowered into the pool well over two hundred feet when they finally ran out of rope. Realizing they were incapable of retrieving the treasure from what they now referred to as a bottomless pit, the discouraged party abandoned any further attempts to recover it.

The Grand Rock Water Hole is located just a few miles northwest of Carrizo Springs in Dimmit County and not far from Pena Creek. A recovery party that visited the spring during the early 1960s found it dry. Two members of the party rappelled over two hundred feet into the hole but never reached bottom.

A second group, this one composed of professional treasure hunters, arrived at the Grand Rock Water Hole in 1973 and reported it was filled to the top with silt and debris as a result of recent flooding of Pena Creek.

In 1990, according to another treasure hunter, the spring was filled once again with cool, clear water and the bottom could not be seen.

Most who have researched the incident of the Grand Rock Water Hole are convinced a fortune in gold and silver—one worth several million dollars—still lies deep in the bottom of the spring.

Campfire Gold

The old Brownsville to San Antonio Road stayed busy with commercial traffic and travelers following the Civil War. In San Antonio and several of its neighboring communities, the growing population generated a demand for goods, many of which came from Mexico.

Sam Weatherly saw potential in this rapidly expanding trade and developed an import business—he would ascertain what the San Antonio merchants needed, travel to Mexico to purchase as much as he could, arrange for transportation, and then sit back and count his profits.

It was a warm and sunny day in early April 1879 as Weatherly rode alone on the road to Brownsville. He had just passed through the small town of Beeville where he spent most of the day looking for new customers. He wanted to get in a few more miles of travel, and there were still about two hours of daylight left when he rode out of town toward the south. In his saddlebags was close to $20,000 in gold coins, some of which he would use to purchase goods in Mexico, and the rest would be used to open an account at the bank at Brownsville.

Only a few minutes out of town, Weatherly's horse threw a shoe, and what little daylight remained was used up as he struggled to reattach it. When he was certain the shoe was secure on the animal's hoof, there was just barely enough twilight left to allow him to search for a decent campsite. Minutes later Weatherly found one in a nearby copse of trees about forty yards from the trail, and the stone-ringed fire pit he found there suggested it had been used many times before by other travelers.

After unsaddling and hobbling his horse in a nearby meadow, Weatherly carried his saddlebags over to the campsite and placed them next to the fire pit, built a small fire to ward off the growing chill of the evening, smoked his pipe for a while, and then wrapped himself in his sougan and fell asleep.

A few minutes before dawn of the following morning, Weatherly was awakened by a noise. Listening intently, he heard horses approaching, the squeak of leather saddles, and the low voices of men. Sitting up in his bedroll, Weatherly peered into the fading darkness and saw that three riders had left the trail and were approaching his campsite. He also saw they each carried drawn revolvers.

Weatherly's mind raced as he tried to decide what to do. It appeared as though the strangers were intent on robbing him, perhaps even killing him. Using a flat rock, he hastily scooped out a depression in the middle of the fire pit, dumped the gold coins from his saddlebags into the excavation, and covered them over with coals that were still smoldering. As quietly as he could, he ran to his horse, unhobbled it, and without bothering to saddle it, leaped upon its back. As he dug his heels into his mount's ribs and fled southward, he heard angry shouts from the strangers—they had spotted him and set off in pursuit.

Two miles down the trail, the three riders overtook Weatherly and forced him to a halt. They told him they were aware he was carrying a great deal of gold and ordered him to turn it over immediately. When Weatherly denied he had any gold, one of the bandits pulled him from his horse, threw him to the ground, and kicked him several times in the ribs.

The rising sun was just beginning to illuminate the center of the copse of trees when the three strangers and Weatherly returned. Weatherly had been tied to his horse, which was led by one of the intruders. After dismounting and tying the businessman to a nearby tree, the outlaws searched the campsite but found little more than Weatherly's saddle, bedroll, and empty saddlebags. As two of the men searched just beyond the perimeter of the site, the third fanned the coals into flames and, using Weatherly's pot, began making coffee. For the next hour the three men sat around the fire drinking coffee as they pondered what to do.

Two of the highwaymen were Mexicans; the third had the look of a half-breed and went by the name Felix. One of the Mexicans approached Weatherly and asked once again where the gold was hidden. When he received no answer, the Mexican punched the bound man in the stomach. Repeated questions were likewise ignored, and the outlaw ultimately inflicted a severe beating on the businessman.

Convinced Weatherly would tell them where the gold was in time, they decided to simply take him with them until such time as he was willing to provide the information. They tied him once again to his horse, remounted, and rode toward the south, leading their prisoner.

Several days later the three outlaws and their captive crossed the Rio Grande into Mexico. Each night Weatherly was bound and beaten as the bandits continued to try to learn the hiding place of his gold. Once the bottoms of his feet were burned with hot coals, but no amount of pain and suffering could get him to talk.

For seven long months Weatherly's ordeal continued. His ration of food was always small, and he lost a great deal of weight. Little more than skin and bones and with his uncut and unkempt hair and beard, Weatherly was indeed a pathetic figure.

One evening at a campsite somewhere in the Mexican state of Coahuila, Weatherly passed out from a combination of fatigue and hunger. That night the bandits did not even bother to tie him up.

It would prove to be a mistake. Weatherly awoke about three hours before dawn. Realizing he was unbound, he listened closely to the snoring of his three captors. Convinced they were all sound asleep, he quietly crawled away. Though weakened, his newfound freedom provided him with an energy that helped him cover several miles before the sun came up. Knowing that the bandits would search for him, Weatherly found a hiding place among the rocks in the foothills of an old and weathered granite mountain range. Lying in the cool shade of a low overhang and sufficiently hidden from the trail, Weatherly nervously listened for the sounds of pursuit.

They were not long in coming. The three outlaws rode slowly among the rocks in search of their former captive. Luck, however, was with Weatherly as the three passed by without suspecting he lay hidden only a few yards away. He remained in his hiding place all day, coming out only when it was dark. By the light of the waxing moon, he started walking northeastward, a direction he knew would take him back to Texas.

Along the way, Weatherly subsisted on berries, prickly pear, wild onions, and the nuts he picked from a tree he could not identify. Fortunately he passed springs and streams and seldom lacked for water. Weeks later it was a ragged-looking Sam Weatherly who finally crossed the Rio Grande at a point a few miles upstream from McAllen in Hidalgo County.

The following day Weatherly was found by a rancher who took him in and fed and cared for him for two weeks. Weatherly told the rancher his amazing tale of captivity and escape and said now that he was back the first thing he wanted to do was retrieve the gold from its hiding place in the fire pit south of Beeville.

Several months were to pass before Weatherly was able to return to Beeville. When he finally arrived at the growing community in the summer of 1880, he wasted little time in riding out to the campsite where the gold was hidden. He stopped at the first copse of trees he encountered and eagerly searched for the fire pit, but there was none.

Confused, Weatherly rode back out to the road to get his bearings. From where he sat upon his horse, he saw several other copses. He examined each of these but still found nothing. As he rode up and down the trail, he found even more, and they all looked similar. In some of these he found evidence of camping, and in two others he found fire pits, but unfortunately neither was the one he remembered, the one in which he cached his gold.

For weeks Weatherly remained in the Beeville area exploring the various copses, but he could never determine which was the one in which he camped that long ago evening. Frustrated and confused, Weatherly eventually abandoned his search and returned to San Antonio to try to reestablish his business. For the next several years as he traveled the Brownsville-San Antonio Road, Weatherly would pause to search once again for the firepit but could never find it. He passed away in 1891.

In 1925 a bank in San Antonio placed on display in their front window a curious item—eight twenty-dollar gold pieces were fused together into one misshapen lump.

A card lying next to the gold was neatly typed with the message, "Found in a fire pit south of Beeville, Texas, on June 8, 1918, by Wendell Baker." Baker was one of the bank's customers.

Did Baker accidentally find Sam Weatherly's hiding place? If so, were these eight coins the only ones he retrieved? It seems likely they were, for Baker was not a man of means and in 1925 had a bank balance of less than thirty dollars.

Logic dictates that Sam Weatherly's hasty gold cache still lies only a few inches below the ground south of Beeville in some long forgotten copse of trees. Perhaps, like Baker, someone may pass this way and eventually find the rest of the gold. The cache, according to coin collectors, could be worth between $250,000 and $500,000.

Big Thicket Treasure Chest Found and Lost

The Trinity River bottoms in East Texas's Liberty County were a treacherous place to be during the 1930s. Though only about fifty miles from Houston, the region was surprisingly remote and unpopulated save for a few poor families. Not far away can be found the Big Thicket, a near-impenetrable swamp that is home to water moccasins, rattlesnakes, alligators, and wild hogs. For generations the Big Thicket had been a favorite hideout for outlaws and renegade Indians.

Floods ravaged the Trinity River bottoms from time to time, washing away homes and farms. Even today things are little changed.

In 1932 a long lost Spanish chest filled with treasure was found in the Trinity River bottoms by a sharecropper, the contents examined, and within minutes, lost again. Subsequent efforts to raise the chest have failed, and it is estimated nearly two million dollars' worth of gold and jewels lie at the bottom of a Big Thicket swamp today.

The tale of the lost chest filled with gold and jewels has been around for a long time. The story goes that a small caravan of Spaniards became lost in the region of the Big Thicket while traveling north to some unknown destination. The chest, it was said, was made of wood, bound with copper straps, and contained golden church articles such as chalices, crosses, and ornaments. Precision handmade knives with jewel-inlaid handles were also reported to be inside the chest. Details are sketchy regarding this treasure, and there are precious few directions. And the Big Thicket is a huge, forbidding place. In spite of all these obstacles, this treasure chest was found, opened, and immediately lost.

In 1932 a black sharecropper named Ben Barney was hunting rabbits near the edge of the swamp. As he was walking along a portion of the shore that ringed a large pond of black water, he found two wooden poles, each about three feet long. The poles had been cut from some kind of hard wood and fitted with leather handles at one end. Barney picked up the poles and carried them to his shack just off the dirt road that snaked its way to the small community of Kenniflick. He showed the poles to his wife, Ella, and then placed them in the lean-to behind the house.

Approximately three weeks later, two strangers arrived at the Barney home. The two men were priests, traveling on foot, and one of them carried a pole similar to the ones earlier found by Barney. Ben was working in the field about a mile away when Ella answered the door. Without making introductions, one of the priests held up the pole and asked Ella if she had ever seen one like it. She claimed she had not, then told the two men she was busy and they would have to leave.

Once the two strangers were out of sight, Ella ran to the field and told Ben about the visit. The two returned to the shack and spent the rest of the day peering out the windows, fearing the strangers might return and do them harm.

The next morning Ben, taking the two poles, ran to the home of a neighbor, Jed Murphy, and told him what had occurred. Murphy asked Ben to take him to the place where he found the poles, and together they returned to the edge of the swamp. After looking around, Murphy found several places where someone had spent considerable time walking along the shore and apparently using the poles as probes. They were obviously looking for something in the dark waters of the pond. Returning to his house, Murphy located a seven-foot metal rod and walked back to the swamp. Using the longer rod, Murphy jammed it again and again into the marshy ground. After about ten minutes and while he was standing at the edge of a low, steep bank, one of his thrusts made contact with a solid object. Repeated jabs with the pole yielded a sound that indicated the object struck was wooden.

After marking the location, the two returned to Murphy's house and gathered up a block and tackle rig, two shovels, a pickaxe, and double-harnessed two mules. After returning to the swamp with the mules, Murphy held one end of the chain while Barney dragged the other into the deep water. After struggling for nearly an hour with the muck and sediments, Barney finally succeeded in getting the chain around the object. After he was safely back onto the shore, Murphy looped the chain over a makeshift tripod and hitched it to the mules' harness. With a command to the two animals, he started them pulling. Grudgingly, the swamp yielded the mysterious object—ten minutes later a large, swamp-stained, and partially rotted wooden chest rested just on the edge of the steep slope.

The lid to the chest was fastened with a very heavy and ancient-looking lock. Using the pickaxe, Murphy beat the lock but it would not give. With the chain still wrapped around the chest, he used the pickaxe to pry the lid back. Finally a portion of the metal top was peeled back, and the two men stood in amazement as they gazed down at the contents. Lying before them was an assortment of gold chalices, crosses, rings, and coins. The handles of several knives appeared to be inlaid with diamonds and emeralds.

Murphy and Barney stared dumfounded at this amazing treasure. Just as they were beginning to realize they might be very rich men, one leg of the tripod snapped and the chest, already teetering on the edge of the slope, slid backwards until it was partially submerged in the dark water. Murphy whipped the mules, and just as the animals started pulling the chest back up, the chain slipped off and the treasure slipped back into the depths of the pond.

Once again Barney went into the murky water with the chain but was unable to reattach it to the chest. At one point he became mired in quicksand as he was crawling back onto shore, and with great difficulty Murphy finally pulled him free. The two men decided to quit for the day and make another attempt at raising the chest the next morning.

Murphy and Barney arrived at the site around 9:30 a.m. the following day. As Barney was stripping off his overalls prior to entering the pond, he was bitten on the calf by a water moccasin. Murphy treated the bite as best he could and half carried, half dragged the weakened man back to his home. By midafternoon Barney's calf had swollen to twice its normal size, and he was in great pain. Borrowing a wagon from another neighbor, Ella carried him to Kenniflick where he was placed in the care of a "root doctor." For four days and nights, Barney remained confined to his bed, but eventually the swelling went down and he was able to go home. The damage from the snakebite was so severe, however, that he was forced to walk with a cane.

It was several weeks later when Barney and Murphy were able to return to the swamp where they had brought the treasure chest to the surface. They made several more attempts to loop a chain around the chest, but Barney, weakened by his ordeal, was forced to quit. Several more attempts during the ensuing month likewise ended in failure.

Nine years later Ben and Ella Barney, both in their sixties, died. He and Ella lived their entire life in poverty and, though he was once only moments away from being one of the wealthiest men in East Texas, he died the same way.

Several weeks following the death of his friend, Jed Murphy went to the man who owned the land on which the swamp was located. He told him about the wooden chest filled with treasure and described in detail the cups and crosses made of gold. Intrigued, the landowner brought in a team of professional salvors. After being led by Murphy to the place where the chest was first brought up, they began probing the bottom of the dark pool with long metal rods. They found nothing.

One of the salvors, donning SCUBA gear, slipped underwater in search of the chest. Though he did not find it, he did find part of its contents: He came back to the surface with a small cross made of gold. He said it had been caught by an underwater root. He also announced that the bottom was mostly composed of quicksand, and he presumed the chest must have sunk into it. The salvors ultimately informed the landowner that recovering the chest was impossible, and they abandoned the site.

Not willing to give up, the landowner, utilizing some connections he maintained with the huge Texas oil and gas industry, secured the services of two mining engineers. Using sophisticated scanning equipment, the engineers located the presence of a large metal object several feet below the bottom of the pond. To reach it, they advised, the pond must be drained and the bottom dredged up. Both jobs, they admitted, would be quite costly and nearly impossible. The landowner mulled this information over for several weeks and finally abandoned all hope of ever retrieving the chest.

After conducting some research on the tale of the lost Spanish treasure chest and determining that its contents could be worth as much as two million dollars, two professional treasure hunters from England arrived in the area to examine the prospects of recovery. They spent a week camping along the shore of the swamp. During that time they dove into the dark waters of the pool and employed a number of different electronic sensors. They also contended with thick swarms of mosquitoes, an abundance of poisonous snakes, treacherous quicksand, torrential rainstorms, and hostile residents. In the end they acknowledged that, indeed, there was a treasure chest several feet below the bottom of the swamp but that retrieving it would be a dangerous as well as costly undertaking.

The gold- and jewel-filled Spanish treasure chest is still there, still submerged beneath the black waters of the swamp and buried deep in the soft bottom soil. It continues to lure the curious and the hopeful to the site, still tempts treasure hunters. All it is waiting for is for someone to figure out a way to retrieve it.

Bibliography

A great portion of the information contained in this book was derived from researching old diaries, journals, and period newspapers. Where possible, much of this was followed up by interviews with a variety of people, including many who were descended from some of the principle figures in these stories as well as a number of people who spent considerable searching for one or more of these lost mines and buried treasures. Additionally, each of these tales was followed up with organized expeditions to the specific regions to identify landmarks and ferret out leads pertinent to the actual treasures.

Below, the reader will find a selected bibliography, which includes some of the resources that proved to be among the most useful to the researchers.

Cremony, John C. *Life Among the Apaches*. Lincoln: University of Nebraska Press, 1983. First published in 1868.

Dearen, Patrick. *Castle Gap and the Pecos Frontier*. Fort Worth: Texas Christian University Press, 1988.

Dobie, J. Frank. *Coronado's Children*. New York: The Literary Guild of America, 1931.

Jameson, W. C. *Tales of the Guadalupe Mountains*. Plano, Texas: Republic of Texas Press, 1995.

_______. *The Guadalupe Mountains: Island in the Desert*. El Paso, Texas: Texas Western Press, 1994.

_______. "The Treasure of the Bottomless Pit." *Lost Treasure* (August 1992): 45, 47, 49.

_______. "Jean Laffite's Lost Treasure." *Lost Treasure*. (August 1992): 39-41.

_______. *Buried Treasures of Texas*. Little Rock, Arkansas: August House Publishers, 1991.

_______. "The Lost Treasure of Barkentine Creek." *Lost Treasure* (June 1991): 23-24, 26.

_______. *Buried Treasures of the American Southwest*. Little Rock, Arkansas: August House Publishers, 1989.

_______. "The Lost Goatherder's Treasure, Part I." *Lost Treasure* (March 1991): 14-17.

_______. "The Lost Goatherder's Treasure, Part II." *Lost Treasure* (April 1991): 24-27.

_______. "The Juniper Spring Treasure." *Treasure* (March 1985): 8-10.

_______. "The Sad Saga of Rolth Sublett, Treasure Hunter." *Frontier Times* (August 1985): 60-62.

Miles, Elton. *More Tales of the Big Bend*. College Station, Texas: Texas A&M University Press, 1988.

_______. *Tales of the Big Bend*. College Station, Texas: Texas A&M University Press, 1976.

Rascoe, Jesse Ed. *The Golden Crescent*. Toyahville, Texas: Frontier Book Company, 1962.

Rosenthal, C.M. "The Priest's Gold," *True Treasure*. Winter 1967, pp. 26-30.

Rowen, Will. "Pancho Villa's Texas Treasure," *Lost Treasure*. July 1979, pp. 26-29.

Townsend, Tom. *Texas Treasure Coast*. Austin, Texas: Eakin Press, 1979.

Wattson, Mrs. Ben. *Hidden Gold of the Guadalupes*. El Paso: Tumbleweed Press, 1966.

Williams, Carlos. "The Immigrants' Swan Song: Sand Dunes Treasure." *Treasure Cache*. Grove, Oklahoma: Lost Treasure, Inc. (1994): 49-51.

Wilson, Steve. "Lost Golden Ledge of Eagles." *Golden West*. NPI.

Index